DAILY LIFE AS
SPIRITUAL EXERCISE

by the same author

In English

Hara, the Vital Centre of Man: Allen and Unwin
The Japanese Cult of Tranquillity: Ryder and Co.

In German

Durchbruch zum Wesen (Breakthrough to Being)
Im Zeichen der Grossen Erfahrung (The Great Experience)
Erlebnis und Wandlung (Experience and Transformation)
Der Mensch im Spiegel der Hand (Man Mirrored in His Hand)
Japan und die Kultur der Stille
Hara
Zen und Wir (Zen and Ourselves)
Die Wunderbare Katze (The Wonderful Cat and other Zen Texts)

In preparation

Personale Therapie (Personal Therapy)

DAILY LIFE AS SPIRITUAL EXERCISE:

The Way of Transformation

by Karlfried, Graf von Dürckheim

Translated by Ruth Lewinnek
and P. L. Travers

PERENNIAL LIBRARY
Harper & Row, Publishers
New York, Evanston, San Francisco, London

This book was originally published in England under the title *The Way of Transformation: Daily Life as Spiritual Exercise* by George Allen & Unwin, Ltd. It is here reprinted by arrangement.

First PERENNIAL LIBRARY edition published 1972.

STANDARD BOOK NUMBER: 06-0802626

Designed by Yvette A. Vogel

CONTENTS

Everyday Life as Practice 1

Healing Power and Gesture 47

The Wheel of Transformation 73

    1. *Critical Awareness* 75
    2. *Letting Go* 84
    3. *Becoming One with the Ground* 90
    4. *New-Becoming* 104
    5. *Proof and Practice in Everyday Life* 121

Appendix 135

Everyday Life as Practice

It is a truism that all work, all art and all professional activity require practice if they are to succeed. This we accept, and in order that we may establish ourselves in the world, it is obvious that we must be at pains in all our vocations, avocations and transactions to practise and assimilate experience. We do not sufficiently realize, however, that the success of man's most important task—infinitely more essential than any of his arts or professions—also depends upon practice.

The destiny of everything that lives is that it should unfold its own nature to its maximum possibility. Man is no exception. But he cannot—as a tree or a flower does—fulfil this destiny automatically. He is only permitted to become fully what he is intended to be when he takes himself in hand, works on himself, and practises ceaselessly to reach perfection. Here we must ask ourselves the question —what is man's most important task? It is none other than himself, the making of himself into a "true man."

The readers may well ask—what purpose do study, practice and the collecting and assimilation of experience serve when one is oneself the task at hand? What steps are necessary in order to achieve this? What kind of experiences need to be thought of as essential and therefore to be integrated? In what does practice consist? What are the prerequisites of success?

*

Let us first consider what we conceive of as a worldly task. It is evident that the prerequisites for this to succeed are: a mind completely at the service of the work in hand, a tenacious will, a capacity to assimilate the necessary experience, the efficient development of relevant talents and their proper techniques and, in addition to all this, the ability to achieve a continuous adaptation to the exterior world. As a result of such labors it is possible for man to bring to fruition that mastery through which worldly success is assured.

However, our inner task, if it is to prosper, must be the fruit of a human being's complete maturity in all his aspects. And the prerequisites of maturity consist first in the breaking down of the small "I"—that is to say, the "I" that rejects pain, is full of fears and is solely concerned with the things of the world. Then must come the perception, the acceptance and unfolding of our inborn, transcendental Essence and, together with all this, the relinquishing of standpoints and positions which obstruct it. There must be an attitude of earnest acquiescence to such experiences as will lead to its perception, the attainment of a personal posture which corresponds to the inner essence and, throughout this difficult work, an absolute devotion to one's progress along the Inner Way. By these means a man is led towards inner mastery. An all-encompassing attitude will thus have been achieved which will enable the process of inner growth and ripening to continue. Such a life attitude is the only one by which a man may fulfil his own law and destiny, which is to become a Person, a human being who is at once transparent to the Divine Being within him and able to express it in his life in the world. Our inner task, unlike our work in the world, does not revolve around external aptitudes but has as its focal point the transformation of the being a man into an expression of Divine Being.[1]

If and when it happens that the inner task prospers, it must not therefore be assumed that a man knows or possesses more than before. What happens is that he *is* more. A change has taken place. Behind the worldly achievements visible to any observer there is the inwardly transformed man who, very possibly, may be visible to nobody. But just as all valid capability in the world requires a certain level of human development, so also does the transformation which leads to inner maturity require that a man should conscientiously and responsibly labor at the tasks *set by the world*. Inner and outer work are not mutually exclusive. On the contrary, we are invaded, as it were, from morning to night, both by our inner being as well as by the threatening exterior world for which we are responsible. The field of our ceaseless effort to reconcile both sides is none other than our ordinary daily life.

*

Our achievements in the world may be said to be perfect only in so far as they consummate and realize the idea that is their inner meaning. Whether it be a

[1] See Appendix.

work of art or a system of philosophy, the building of a house or the setting up of some organization, factory or technical apparatus, it possesses a valid form (*gültige Gestalt*) only when each of its parts accomplishes the underlying purpose of the whole. The same applies to the inner task. Man, himself, is one of life's forms (*Lebensgestalt*) but he, too, is valid and can endure only in so far as he fulfils, as a whole and in the unifying of all his forces, what he is fundamentally intended to be.

But what is man? What is he intended to be? Intrinsically—that is to say, in his essence—man is an aspect of Divine Being, whose purpose is to manifest itself in him and through him. Just as flowers and animals in their own way reveal Divine Being, so also must man reveal it in the way of man. He can only be "right" when his life-form, with all its forces, fulfils the destiny that dwells within him as his essential being.

For each of us, our essential being is the medium through which we participate in Divine Being. It may also be said to be the means by which Divine Being strives through us to manifest itself in the world—not as something apart from the world, a mere spiritual inwardness, but as a bodily presence. It follows that our human inner growth is necessarily accomplished within the framework of our worldly destiny, in the midst of ordinary life and the carrying out of daily tasks.

*

Man's destiny is to bear witness, in his own special way, to Divine Being—which is to say, consciously

and freely. Flowers and animals fulfil theirs of necessity and unconsciously—limited only by whatever
external conditions may hinder them from becoming
what, in their essence, they are. Man, however, unlike flowers and animals, has been endowed with
consciousness; and it is by means of this quality that,
as an "I," he is able to stand and confront the world.
Because of this he becomes, at least to some extent,
independent and therefore responsible for his own
development. Herein lies his great chance. But here,
too, let us not forget, lies danger. For between this
chance and this danger he may miss the mark.

*

The "oneness" of that Life which is beyond space
and time breaks apart in man's ego-consciousness into
two poles—one, the *historical world* which, being
subject to laws of space and time, can be understood
and mastered; and two, *Transcendental Being*, which
cannot be encompassed by space and time and is
beyond all ordinary understanding. Thus, inevitably,
man experiences himself as a being standing between
heaven and earth. He lives in a state of tension between two realities—the world, which confines him
to latitude, longitude and duration, endangers his existence, tempts him with hopes of happiness, and demands his obedience to its laws; and the supernatural
state of Being hidden within him and towards which,
even unknowingly, he yearns. Ceaselessly striving to
find the light, this Being calls him forth beyond the
laws and boundaries of his little personal life, to the
service of Greater Life.

*

It is essential that each person, in order to be able to endure his fate in the world and to master his own life, should amass knowledge of himself and of the way he exists as a temporal being. To become a "whole" man, however, he must also come to have knowledge of the one he is in his divine essence, i.e. of his essential being, whose needs must also be satisfied. Only when he has acquired this knowledge will he be free to satisfy the demands of his inner being. However, the law of his development is such that in the beginning it is predominantly the consciousness, which serves to establish his place in the world, that is unfolded. This one-sided view of the world, which is dominated by the will to survive and the wish that life should have meaning and provide satisfaction, veils the Divine Being present in his own essential being. Thus man readily loses sight of the deeper significance of his existence which is, in effect, to manifest Divine Being within this life. Ultimately, however, he cannot escape from the fact that his wholeness and welfare depend on his fulfilling his inner mission and living from his essence. If, consciously or unconsciously, he concerns himself solely with his work in the world and exercises and establishes his world-ego alone, he is bound in the end to fall prey to a specifically human unhappiness. And this happens whether or not his personal virtues and achievements correspond to the values and expectations of the world.

*

The more man has succeeded in adapting himself to the world by mastering external life and the more he assumes himself to be living blamelessly, the less is he able to understand the inevitable and, to him, inexplicable suffering which is brought about by his estrangement from his inner essence. This suffering, coming as it does from his still unfulfilled inner being, is of a quality quite different from any unhappiness the "I" may experience in the world. Only that pain which arises from the depths of himself can finally compel a man to look inwards and bring him to the realization that he has lost touch with his own essential inner being. In such a condition he may perhaps recall times when he sensed something beckoning to him from within and turned his back upon it; moments when conscience summoned him, and he was not able to listen. Thus he is faced with a decision—either to stifle the inner voice once again and continue in his old ways, or to make a new start based on what he has learned from within.

*

Once, however, a man has so awakened to his essential being that he can no longer ignore its claims, his life from that moment is governed by a new and continuous state of tension. Inevitably, now, he will be oppressed by the contradiction between the needs, duties and temptations of the world and the call of his inner nature. On the one hand the world will continue to exact its requirements without reference to the inner voice and, on the other, without reference to the world's voice, inner being will make its insistent demand.

Thus we are torn between the two aspects of our human existence. And yet our life in the world, as well as our participation in Transcendent Being, are but two poles of the *one self* that is ever striving within us towards realization. It is through the development of this self that the oneness of life is realized in the human form. And, given these conditions, if we are upright within it will inevitably mean that we are upright without. For this reason it is essential to discover in ourselves an attitude—even a physical posture—in which we can be open and submissive to the demands of our inner being while at the same time allowing this inner being to become visible and effective in the midst of our life in the world. And for this to happen we must so transform our ordinary daily life that every action is an opportunity for inner work. Our very efforts towards worldly efficiency may, under these conditions, become the means of inner practice (*exercitium*).

*

The purpose of all living things, man among them, is to manifest the divine in the world. Man's distinctive virtue lies in the fact that the great, the Divine Life becomes in him, "conscious of itself." In the freedom of conscious life—as opposed to mechanical life—the divine can shine forth and take form. Therefore the true man is he who, in freedom and with clear awareness, embodies and reveals the Divine Being within himself. The vividness of his inner life, the radiance of his being and the benign effect of his actions will all give evidence of his inner state. Such a state enables Divine Being, in all its plenitude, inner order and unity, to appear. It does not, however, develop along a straight line for it must first pass through a stage of existence in which the "I" is determined solely by the world, thus preventing a man's awareness of his essential being. This so-called "natural" manner of existing, whose center of consciousness is the worldly-ego, must of necessity be broken down, for this ego is not concerned with the need to ripe in accordance with inner being but is solely preoccupied with maintaining its secure position and competent functioning in the world.

*

Man survives in the world by means of a certain kind of consciousness which, by its crystallized concepts, organizes life into a succession of rigid structures. Through these he perceives the world rationally, masters it systematically, and orders it according to fixed values. It is just this form of existence that veils his awareness of Divine Being. At the center of this way of life stands the ego which, by its very nature, is solely occupied in establishing and classifying "facts," and around it revolve continually its own worldly desires. By means of this ego, man separates himself from his unconscious connection with the oneness of life, and assumes in the world an attitude of independence and self-assertion. As long as the ego takes up uncompromising attitudes and irrevocably fixed points of view, so will the oneness of life break up into the opposites of I-World/World-I and essential being. The ego's separation from the primal core of life and its determined orientation towards the world, bring all growth, all true becoming, all transformation, to a disastrous standstill. It is true that man has need of an ego that is capable of objec-

tively mastering the world. But only when he succeeds in evolving a form of existence in which his
world-ego is sustained (though in such a way that it
remains at the same time transparent to Divine Being
within him) can he become a "true" man—a
"person" in the ultimate sense, through whose life
Greater Life resounds (*personare* = to sound
through). To attain this state of mind requires ceaseless practice. In other words, every passing moment
of the daily round must be used as an opportunity
for such practice.

When one speaks of the daily round the phrase
carries overtones of meaning and experience which
distinguish it from the holiday. Compared with a
holiday the ordinary day may appear montonous and
grey. It is dull in the sense that one is used to it, the
common round of activities always has an unchanging sameness. By this endless repetition the normal
day threatens to mechanize us. It is a treadmill of repeated movements and actions, whereas the holiday
stands for something unique, joyous, unusual. The
ordinary day deadens, whereas the holiday refreshes.
If one contrasts the freedom Sunday brings us, the
ordinary day feels constrained and rigid; it stands for
labor, for a round of doing, in contrast to the leisure
promised by Sunday. Its sobriety is far from the fes-
tivity which a holiday brings. The ordinary day de-
vours us; on holiday we find ourselves again. The ordinary day draws us outwards, Sunday enables us to
be inwardly free. The ordinary day is ruled by the
overbearing world which compels us to be efficient
machines, but Sunday belongs to our inner, nonme-
chanical selves.

Need this be so? It must be so, alas, as long as man

is engulfed by his world-ego, and so engrossed by worldly undertakings that they completely overshadow his inward self. There is, however, a way out. It needs just a single moment of insight, a flash of understanding of our situation—to make everything change. Such a moment, such a flash will light up not only our outer actions—particularly those most repetitive and familiar—but our inner experience as well.

It is proverbial that in whatever we do we assume a certain attitude. "What" we do belongs to the world. In the "how," the *way* we do it, we infallibly reveal ourselves whether our attitude is right or wrong in harmony with the inner law or in contradiction to it, in accordance with our right form or opposed to it, open to Divine Being or closed to it. And here we must pose the question: what *is* our right "form"? It is none other than that in which we are transparent to Divine Being. And to be transparent means that we are able to experience Divine Being in ourselves and to reveal it in the world.

*

Let us suppose, for instance, that a letter has to be posted in a pillar-box a hundred yards away. If the mouth of the pillar-box is all we see in the mind's eye, then the hundred strides we take towards it are wasted. But if a man is on the Way as a human being and filled with the sense of all that this implies, then even this short walk, providing he maintains the right attitude and posture, can serve to put him to rights and renew himself from the well of inner essence.

The same can be true of any daily activity. The

more we have mastered some relevant technique, and the smaller the amount of attention needed to perform the task satisfactorily, the more easily may the emphasis be transferred from the exterior to the interior. Whether in the kitchen or working at an assembly-belt, at the typewriter or in the garden, talking, writing, sitting, walking or standing, dealing with some daily occurrence, or conversing with someone dear to us—whatever it may be, we can approach it "from within" and use it as an opportunity for the practice of becoming a true man. Naturally, this is possible only when we are able to grasp the real meaning of life and become responsible towards it. It is essential to realize that man is not committed merely to comprehending and mastering the external world. He is first and foremost committed to the Inner Way. When this is understood the truth of the old Japanese adage becomes clear: "For something to acquire religious significance, two conditions alone are necessary: it must be simple, and it must be repetitive." [2]

What does this word repetitive signify here? It can happen that daily tasks, by their very familiarity, serve to free us from the grip of the ego and its quenchless thirst for success. They can also help to make us independent of the world's approval, and open for us the inward way. But this is true not only of familiar tasks. Even the practice and repeated effort needed to master something new can be put to the service of the inner work. In everything one does

---

[2] See Dürckheim, "Die geistige Überwindung des Mechanischen" in *Durchbruch zum Wesen* (Breakthrough to Being), Huber, Stuttg. Bern.

it is possible to foster and maintain a state of being
which reflects our true destiny. When this possibility
is actualized the ordinary day is no longer ordinary.
It can even become an adventure of the spirit. In
such a case the eternal repetitions in the exterior
world are transformed into an endlessly flowing and
circulating inner fountain. Indeed, once repetition is
established it will be found that our very habits can
be the occasion for inner work. They enable us to
make new discoveries and show us that even from
the most mechanical actions there may issue forth
that creative power which transforms man from
within.

*

By dwelling exclusively in his I-world consciousness—which develops out of, and lives by, crystallized concepts and values—a man loses touch with
Transcendental Being. But in his essential being he is
and forever remains a "mode" of this Being. Without
ceasing, it strives to manifest itself in and through
man, as a threefold process: as plenitude bringing
joy, as order bringing meaning, as oneness manifesting itself through love—all three being invulnerable
to the world. Thus one becomes aware that Being,
with its driving force present and alive in our essence, is the result neither of wishful thinking, nor of
religious belief, but is the most real of all realities.
Only for the man who is entrenched in his worldly
positions, rational concepts and rigid habits does
Being remain an abstract idea. For him who has lost
touch with the core of life it is, at most, a subject for
metaphysical speculation or pious credulity. But for

one who breaks through the mists of rationalization, it becomes personal experience. All religions, concepts and images are, in their original meaning, interpretations of this deepest of all experiences.

When, however, religion is allowed to become a system of rigidly ordered principles and dogmas—i.e., forced into the confines of a doctrine that is required to be believed—it blocks the possibility of vitally experiencing Divine Being, or alternatively it tries to debase it by labelling it as a "merely subjective" experience. But what does such a phrase connote? In point of fact, it means the existential truth of man as an experiencing and suffering subject—and to this condition he is brought solely through this so-called "merely subjective" inner experience.

The time has come for man to free himself from the idea that only those things which can be rationally grasped and objectively verified are real and binding. With regard to questions of being and the meaning of life he needs to recognize the primal validity of the inner experience which, beyond all doubt and with unequivocal clarity, when it arises within him, brings joy, a sense of responsibility and transformation. The undeniable inner experience is the one thing that can truly touch, move and compel us to follow it. And only by completely experiencing Divine Being can we come to realize that the whole content of the world, as we see it, is itself nothing other than that Being. But for us that Being is clouded; we perceive it through the refracting and distorting prism of the ego—that ego which, being externally orientated, inevitably pins down and classifies everything it observes or experiences in terms of objects and of opposites.

Such a rational ego-view inevitably has the effect of solidifying our consciousness, so that we take as real only that which fits conveniently into its preconceived scheme of things. Under such conditions, man loses awareness of his contact with Being. The only reality he believes in is the one he intellectually understands, that in which Life, under the fixed gaze of his ego, appears to present itself to him. The more man has shut himself off from Divine Being by this attitude, the more is his own inner essence repressed —that inner essence which is our only approach to the revelation of Divine Being. It is the suffering which comes from such repression that makes us ready to give proper weight to those rare moments in which Divine Being blazes up within us, bestowing joy, giving us a sense of freedom and calling us to serve its purpose. But man will only be able truly to value such an experience when he is able to recognize that other suffering which engulfs him when he allows the world-orientated ego to hide him from true Being. Only then will he be able to begin the process of withdrawing his ego to some extent from its world-pursuits. Only then will he become open within, free to sense Greater Life directly, able to hear its voice and thus to transform himself and all that surrounds him.

*

The state of mind and body that blocks any awareness of the presence of Greater Being inevitably produces a special kind of suffering. It is the experiencing of this anguish that forces man to take stock of his life attitudes. This taking stock, if honestly carried out, brings him to so great a pitch of desolation that it forces him to the point where he can perceive, sense and heed the call of the true essence towards which he now feels himself driven. If, in some propitious hour, he has been given this experience of Being, and if, through the failures of his world-ego and the suffering that results from them he has been brought to the great turning point, then at that moment he finds himself on the threshold of something new. Now without fail he must open himself to the mysterious, numinous content of moods and impulses that he can neither understand nor explain, still less translate into concepts, but in which he can, nevertheless, unmistakably sense something from beyond this world beckoning and calling to him. He must come to realize that through such experiences Divine Being, as has already been said, is present in his own being. He must train his faculty of sensory perception to feel and appreciate qualities which are beyond the limits of his ordinary consciousness. His effort now must be to learn how to enable Divine Being to enter his own being and remain within him. And, in addition, he must strive to reach a state where his transparence to Divine Being endures, thus enabling him to draw life from Greater Life, even in his existence as an "I" in the little world. To these ends he needs inevitably to cleanse his unconscious and discard any element, spiritual or physical, that dams up the flow of his essential being. At the same time he must give entrance to and build upon all that is in harmony with this being. Furthermore, he must practise sustaining this state in his everyday life and, in so doing, he will infallibly come to learn that daily life itself is a "field of opportunity."

There is nothing—no event, no activity—that cannot provide a motive for establishing contact with inner being, nothing that cannot be used to bear witness to it. Any situation in life may be the means of furthering the bringing to birth of the "true man." For man is only "true" when he listens inwardly to Being, and at the same time remains in accord with it in his bodily existence in the ordinary world. It is not enough just to turn inwards, to experience Being solely from within. To express it outwardly is even more important, for it is imperative that man's life in the world as well as in his body, should derive from inner essence.

*

Side by side with the rediscovery of Man, which is the distinguishing mark of our age, a new consciousness of the body is becoming increasingly evident. The meaning which we give to our body depends on the extent to which we value our human existence. If we measure ourselves solely by our relation to the daily world that continually oppresses and threatens us with its endless requirements, then our body has no meaning except to make our performance and survival in that world physically possible. The whole point of bodily training is thus narrowed down to the development and maintenance of physical efficiency and good health, the sole purpose of which is to function smoothly at its own level. On the other hand once we understand that the real meaning of human existence is to make manifest in the world the Divine Being embodied within us, then the true significance of bodily training lies in the need to reach a state that makes such manifestation possible. A man may be at the height of his physical powers, however, and yet not be able to achieve this. His "presence" may be, as it were, unbecoming. Indeed, his very effectiveness in the world, in a bodily sense, may actually obstruct his essential being. On the other hand it is possible for a dying man to be physically present in the right way if, serene and trustful, he is able to let go and give himself up, allowing that which is meant by "Death as Transformation" to take place. Divine Being, since it is beyond both living and dying, is revealed with equal clarity in life as in death; and man, at the moment of death—even in his bodily condition—may become transparent to Greater Life. It is thus possible to distinguish three different kinds of physical "conscience." The demands of the first are orientated towards health, the second towards beauty and the third towards transparence to Greater Life.

*

Today, it should be generally recognized that what we call "body" and "psyche" are not two separate entities; rather, they are the two modes by which man, with his rational, divided vision, perceives himself. They are the means whereby man as subject realizes and expresses his personal unity. That which is without is within, and that which is within is without. The unity which encompasses and comprehends this antithesis of body and soul is the Personal Subject. And the only kind of perception with which we can do justice to man as a personal subject is the one that sees him not merely as an objective "thing," but—by participation in his life—learns to value him as a subject and recognize him as a "Thou." Here it is necessary to distinguish between the discriminating "what" consciousness that is concerned with establishing facts and the understanding "who" consciousness which alone can reveal personal truth.

*

As *subject*, we experience ourselves as part of a system which has little relation to the outward world in which all things are rationally ordered and fixed in unchanging positions. From the point of view of this world, we see everything materially—as something outside us. We even turn ourselves and other people into objects, into mere "pieces of the world." In as far as we see ourselves and others in this way, we are obliged, as are all "things," to fit into and function within rigid, worldly—and, on their own level, valid —systems and structures. However, when we relate to life as subject, as Thou, we experience and apply towards it a *personal point of view*. Then everything is seen in terms of the life that we ourselves *are*—that life which, in accordance with its own law, is forever striving towards consciousness, completion, liberation and fulfilment. In this way the world resolves itself into the field of our personal life. It becomes the mirror of all our desires, hopes and fears.

*

Life never stands still. Its order is never rigid, like that of some system established by the world. It is always flexible, always changing, always in the process of becoming. And so it is with man. He is not merely an external form, but a Form-in-Time (*Zeitgestalt*) and therefore forever in the act of becoming. The driving force behind this transforming process derives not only from the promise of possible realization which lies in man's inmost essence, but also from the anguish which arises when obstructions prevent the fulfilment of this promise. It is *promise and pain together* that determine the personal nature of the world.

If a man is to become a Person, however, his inner, mental-spiritual development is not the sole requirement: he needs, too, to develop his body. If we regard and treat the body merely as a thing, a piece of the world, we remain blind to its meaning. Even in his body, man as subject is not a thing, a mere biological organism. But what do we mean by the "body" of a person? It is the mode in which he, as subject, is visibly present to the world. Taken thus, the body, in a human sense, can only be rightly understood and worked on when we continually bear in mind the person who lives as this body, that person who—as a "some-body"—is always renewing himself, always striving to unfold and to *be* in the world, in accordance with his inner law. Thus the body is not only something which man has, but the somebody he is in the world.

The true man is he who is present in the world in the right way. Whether he be dense or transparent, withdrawn into himself or open to life, in form or out of form, centered or without center, he is all these things with his whole being, which is to say, also

with his body! Any and every gesture of a man reveals *him* to the world. And it is clear that when a
gesture is frequently repeated and becomes more and
more habitual, the content it expresses also becomes,
for good or ill, more and more crystallized. Thus the
body, in its posture, its patterns of tension and relaxation, in the rhythm of its breathing and manner of its
movement, is an infallible indication of the point at
which any man has arrived on his way to becoming
a Person. It may reveal how and where he has stuck
fast in his ego and lost himself to the world, or alternatively to what extent he has remained open to his
being and on the Way.

*

Practice on ourselves, in the physical and spiritual
sense, is always of two kinds. It involves both the
pulling-down of everything that stands in the way of
our contact with Divine Being, and the *building-up*
of a "form" which, by remaining accessible to its
inner life, preserves this contact and affirms it in
every activity in the world.

Our union with Divine Being, which is endlessly
striving towards transformation, is obstructed by everything within us that is in any way rigid. From his
earliest years man develops, according to his degree
of consciousness, a way of existing in which, as an I,
he is compelled to do consciously and from choice
what he can no longer do unconsciously and instinctively. Thus, whenever the world threatens him, does
not conform to his ideas, or disappoints him, he begins—even in childhood—to develop patterns of be-

havior adapted to his surroundings (*Passformen*). These can help him to survive in situations that may be dangerous, antagonistic, even loveless, but if they become too rigid, they may well obstruct his contact with his inmost being. In such a case, *original trust* is replaced by a reliance on knowledge, on individual capacity, on possessions and all their false promise of security; *original faith*, which rests on unconscious inner laws and the uninhibited evolution of the inborn image, gives place to reliance on conventional patterns and the effort of keeping up the appearance required by the world; *original union* with *Life* is supplanted by dependence on the love and acceptance of others.

The still unbroken chains of our childhood, our imprisonment in the superficial social life, our falsely conceived adaptation to the world, and the ceaseless longing of our ego for security—all these have as their result a mixture of prejudices, ready-made formulas of behavior, and grandiose ideas of what we would like to be in the world. If such aspects of ourselves become rigid—as is usually the case—the first thing necessary is that they be recognised and, after that, dissolved. Of course, if this process of inner rigidity has gone too far, i.e., has become neurotic, professional help may be needed to effect the dissolution. But the more man himself becomes sensitive to the claims of his essence, and the more ready he is to enter into and accept the truth, the more easily can he—if he is earnestly trying to find himself—recognize the inappropriateness of his one-sided worldly standpoint and become aware of those attitudes that estrange him from Divine Being. Gradually, under such circumstances, he becomes more and more able to transform himself. And it is his daily life and its practice that gives him this opportunity.

*

One indispensable quality of all practice on the Inner Way is the presentiment, indeed the experience, of one's own essential being. Without a living contact with this, all practice is bound to go astray and lead into a blind alley. When this happens, we are left to the mercy of the kind of self-discipline that, at the expense of inner truth, seeks health and virtue only in a worldly context. A man who is blind to his own essential being will interpret the suffering which he experiences as an inevitable result of such a condition, as being caused by the world. He may attribute the decline of his efficiency solely to external elements, or feel perhaps that his distress arises as a corollary of some offense he may conceive himself to have committed against the demands of worldly values, religious dogma or someone in authority. Thus the very virtues—so-called—of his all too rigid, hapless, worldly ego, drive him into still greater alienation from his essence. It cannot be too strongly urged that every attempt to realize the true self is doomed to failure unless the depths within are cleansed first. Only through this contact with his essential being can man become sensitive to what is genuine in him, and ascertain what is needful for his self-realization. Only by the experience of his essential being will he come into contact with the Divine Source of Being. And only when this happens is he able to begin the task of creating for himself a "form" which truly corresponds to his own essential

nature and frees him from all falsely conceived methods of adaptation.

It would be quite impossible for man to make any
statement about the nature of Divine Being, were it
not for identifiable encounters, experiences and revelations which in the abundance of their power, significance and feeling and in the sense of responsibility
they bring, are so utterly different in quality from
his usual experience of the world, that he cannot help
but see them as manifestations of transcendence. Experiences such as these are only possible when man's
hitherto impermeable worldly façade has become po-
rous or, indeed, broken down. It is these experiences
which can initiate the "opening of the Way" (*Wende
zum Wege*). All exercises which serve under transformation, if they are not the result of mere defer-
ence to a barren faith in one or other authority, stand
under the sign of such experience. They bring a
sense of liberation and responsibility by means of
which life is seen to be "here and now" and Being as
inward knowing (*Innesein*). Thus, the first and most
vital practice in everyday life is to learn effectively
to value those moments in which we are touched by
something hitherto undreamt of.

*

It can happen that from time to time we are given
moments of illumination and decision when something moves and stirs in our inmost depths. When
this occurs we feel bound to keep faith with the experience, we cannot but obey, and this despite—or
rather, because—what has happened to us is
inconceivable. It is, in fact, something which our ra-

tional "I" cannot really admit. But, if we are able at times really to accept it, our fear of annihilation is assuaged. A deep contentment comes upon us. We sense that within us there is something indestructible which by its plenitude and power creates a new consciousness of life—and thus our trust is renewed. Or again, there may be occasions when we are thrown into despair by some apparently meaningless paradox in our lives, by the "absurd." But if we are able to accept the incomprehensible, it may happen that in the very midst of our lack of understanding, we will sense within the situation some deeper implication. The reality which we understand with our reasoning mind is suddenly transcended and penetrated by another. The profound meaning of this reality only begins to resound in us when we have come to the end of our rational understanding. Here, at the point where our hitherto impregnable belief in the sense and justice of the world has been shattered, there will dawn for the first time a new, fructifying faith in life and its significance. Or something similar may happen when some blow of fate plunges us into utter desolation. Man is intended to live in dialogue with others: he cannot fulfil his life in isolation. He needs a partner, and strives to find shelter in human communion and love. When it happens that life withholds this, or deprives him of it, he is driven to an extremity in which he feels he must perish. If, now, he can bring himself to submit and accept his situation, he has the chance to break through the tension and rigidity with which his enforced loneliness threatens him. If he can now yield trustfully to his inner truth, he may experience the sensation of being caught up and embraced by an unknown power which, at the same time, makes him receptive and open to the whole of life. Although he is forsaken by the world, something from beyond the world makes him feel at one with everything and somehow under the wing of love.

Such experiences may become the turning point of our life. They are the means by which our inmost essence can speak to us. When our world-ego and its proficiency fail us—that proficiency which until now has seemed to be sufficient for the fulfilling and mastering of life—it is then that the inner voices become audible. It may happen that at the moment when the whole framework of our life is suddenly shattered—the framework which was intended to provide our ego with comfort, meaning and security—the chance of something new may appear. If a man is now able to accept that which, with his "I" alone, he *cannot* accept, he may quite unexpectedly experience the sensation of being borne up by Divine Being, and filled with a new significance and sense of being comforted far beyond the comprehension of his normal consciousness.

*

It is this experience of Greater Life which makes it possible for us to look death in the face. There are occasions when, without any doubt, we sense this Greater Life in life, and these can occur a moment after we have been in despair at the thought of life's meaninglessness. Or again we may suddenly feel at one with all that exists, although we are in fact, in the ordinary world sense, completely forsaken. It is in such moments that the tyranny of our ordinary way of looking at the world is overthrown. These are our auspicious hours. It must be remembered, however, that the power of transformation inherent in such experiences can easily fall into the hands of the devil, who is ever ready and waiting to pounce. It is he who, in the guise of our scepticism, questions the veracity of every experience which cannot be classified and made to fit into our ordinary concepts.

But it is not only those periods when fear and despair drive us beyond the limits of our own strength and wisdom that bring us these paradoxical revelations of Eternal Reality. It is also possible to experience moments of boundless joy, which have a quality of the infinite, and in which the plenitude of Divine Being seems to draw close to us. This kind of joy, it should be remembered, has no relation to any form of worldly pleasure.

However, these radiant, unforgettable, deeply moving and blessed experiences are not the only ones in which we become aware of that Divine Being which is the endlessly effluent source of life. There exist also those less outstanding moments and hours in which, quite unexpectedly, we find ourselves in that unique state where Divine Being may touch us. It may be that all of a sudden we are overwhelmed by an unaccustomed sensation. We seem to be wholly present—completely "there," as it were— even though our attention is not directed towards anything in particular. In a strange way—strange because it is so unfamiliar and unexpected—we feel "rounded" and "complete," and at the same time open in a way that allows inner abundance to well up. Although firmly based on the earth, we seem to float on air. At once absent, as it were, and at the

same time wholly "here," we are full of life, able to
rest completely within ourselves and still be closely
related to all that is; apart from everything and at the
same time within everything, bound to all things yet
clinging to nothing. In some incomprehensible way
we are guided and yet free. Released from objects,
aims and claims, we are poor in the world, yet filled
to the brim with inner power and richness. At such
moments we feel penetrated by something that is at
once most precious and most fragile. Perhaps that is
why instinctively we step delicately and do not try
to look too closely at what is happening within us.
Some ancient knowledge warns us that the warm,
semi-darkness of an awakened heart (*Gemüt*) is
more appropriate here than the cold beam of the rational mind which freezes every aspect of life. It is as
if we heard a voice saying: "See as if you did not see;
hear as if you did not hear; feel as if you did not feel;
have as if you had not!" But then, alas, this miracle
vanishes. Suddenly it disappears. Indeed, for it to disappear, one has only to wonder and ask as "what is
it"? For whatever catches our attention, a thought, an
impression, an event—either from without or within
—immediately changes our state of consciousness.
One moment it is quiet and calm, receptive and un-
questioning, like a chalice open to the pouring in of
wine; and the next it is a sword, sharp and deadly,
which pierces the experience and rationally transfixes
it. Now the world which, a second since, was as if
enchanted and all of a piece with ourselves, falls
again into its usual pattern. How impoverished we
feel, now that we are once again thrown back on our
daily selves and are face to face with the ordinary
round. The experience vanishes like a dream. And

yet it was certainly not a dream! It was the revelation of the true reality which rose up in us because, for a moment, we were open and free from the fetters of our ordinary consciousness. Through such experiences we pass beyond this ordinary consciousness to that which is transcendental. No matter how brief the moments of liberation may be—lasting perhaps a fraction of a second—they make us aware, with compelling evidence, of that life which far surpasses our usual I-World impressions; that life to which in our essence we belong, which in our essence we are, and whose revelation is the goal of all our longing.

*

In order to recognize the reality of this experience of Greater Being, we need nothing more than the "sacred sobriety" of common sense. This, in fact, is that transcendental realism[3] which is neither clouded nor inhibited by preconceived concepts and rational thought structures. This sense of reality permits the unique, unclassifiable quality of the experience to be as it is, accepts and savours it, and because of its very incomprehensibility, is intuitively convinced of its truth. We need practice in order to acquire the possibility of recognizing the quality of this reality. By remaining alert and constantly prepared, we can learn to hear and feel the call of Divine Being in everything that happens to us. For this we need to work diligently in order to become vessels capable of receiving all that is poured into them. And this prac-

[3] See Evola, "Über das Initialische" (About Initiation), *Antaois*, VI, No. 2, 1964.

tice, remember, must not only begin but also end the day. "Learn to live each day to its end in such a way that it becomes a part of Eternity," as the poet says.[4]

We need a new kind of discipline here, one that aims so to develop our inner experience that it is lifted to a higher level. The results of such discipline come not from the sort of practice that is the mere carrying out of specific exercises, but from one that confirms the old saying: "Each moment is the best of all opportunities." Thus all things and all events become the field of practice on our journey along the Inner Way. Moving ever onward and keeping touch with his inmost essence, a man transforms himself, step by step, into a "person," whose transparence to Divine Being makes possible the fulfilment of his human destiny.

*

The first condition of all "correct practice" is that one should understand what that phrase means. It is not the pursuit of such capacities as will fulfil our worldly tasks that is meant here, but rather what may be called "exercitium" on the inner way. The kind of practice which aims at achieving a specific result in the world comes to an end when that result has been obtained. Inner practice, however, begins only after one has achieved technical mastery, in whatever activity it may be—even sport—and consists of endless repetition. Once the technique of some task has been perfected, each repetition of its practice mirrors one's whole inner attitude. Every mistake that is

[4] R. Haldewang, Bonz-Verlags-Buchhandlung, Stuttgart.

made shows up a fault within ourselves. If this is understood, our efforts to accomplish daily tasks can, at
the same time, become inner work on ourselves.[5] In
the same way, practice of right posture and right
breathing can become "exercitium" only after one
has fully mastered the necessary technique. Thus by
means of this continuous repetition, the true, inward
person begins to emerge. In this way the ordinary
daily round can be transformed into "practice."

The second condition for this kind of practice—
continuous exercitium—is that a man must be re-
solved and prepared to lead his life in the world in
accordance with his inner law, and this means living
his life in the service of transcendental reality. He
must at all costs grasp the truth which is that only by
being faithful and committed to this service can he
become whole (*heil*), and, that as long as he continues to live solely for security and happiness in the
world, he will be unable to discover the right way.
When all that we know and do becomes a means for
the revelation of Greater Life in the world, then the
Way may be attained, and the ordinary day itself become one single field of practice. But so long as man
tries to gain something solely for himself through
this practice—whether it be the acquisition of higher
faculties, extraordinary experiences, or tranquillity,
harmony, even his own salvation—he is bound to
miss the Way.

*

[5] See Dürckheim, *Sportliche Leistung—Menschliche Reife*
(Achievement in Sport and Maturity), 1966, Limpert-Verlag,
Frankfurt.

In all true practice we are continually preparing ourselves to attain to the position where life in the service of Greater Being becomes second nature. This service will seem onerous as long as we are still cut off from essential being, and living solely in our world-ego. On the other hand, once a man is really committed to the Way he will shortly have experiences that bring him great joy and a sense of freedom.

The Inner Way, and the practice which enables us to progress along it, has little to do with our usual methods of disciplining ourselves—often against our own nature—to serve unselfishly some worldly cause or community. What is involved here is not merely the discipline by which man masters his instinctive drives and overcomes his nature. Nor is it just a matter of acquiring those virtues through which he subjugates his small ego for the sake of the community and his labours in the world. Neither has it anything to do with the personality which loses face when it does not succeed in the world, and is dishonored when it fails to accomplish its tasks or proves to be an unreliable member of the community. All these are steps that necessarily must be taken on the way towards true becoming and maturity. But man can only be said to have really begun his journey on the Inner Way when the injunctions he blindly obeys are no longer those that conform to the world's scale of values, but rather those which accord with Divine Being within him—that Being which constantly strives to reveal itself in him and through him to the world. Even the most correct and responsible behavior is unable to satisfy these high demands if it is solely the result of an ethical discipline instead of being an expression of man's inner essence. This essence is able to manifest itself freely only when the depths in which it dwells have been cleansed and purified.

*

The most vital element in the practice of the Inner Way consists in learning to become open and perceptive to the inward experience of essential being. For it is out of essential being that Divine Being speaks and calls to us.

All day we are summoned by the world, which demands that we recognize and master it. At the same time Divine Being is perpetually calling us inward. The world requires of us our knowledge and skill, whereas Divine Being insists that we lay aside knowledge and skill for the sake of inner growth. The world expects us to be continually doing whereas Divine Being requires us, quite simply, *to allow the right thing to happen*. The world tries to keep us on a never-ending treadmill, so that we may achieve something that we in our blindness consider permanent. Divine Being demands that while remaining in touch with it, we refrain, at the same time, from becoming attached to anything at all— even though, in so refraining, we may fear to lose our hold on ourselves. The world encourages us to talk and to do interminably; Divine Being requires that we become quiet and act without acting. The world forces us to concentrate on security; Divine Being asks only that we risk ourselves again and again. To the extent that we comprehend it, the world does our bidding; Divine Being reveals itself

only when we accept the incomprehensible. The
power of Divine Being becomes apparent at the moment when we let go of the things that support us in
the world. It is only when we are able to relinquish
that which makes us rich in the world that we are
enriched and transformed by Divine Being.

*

Practice in daily life has many aspects. It requires
that we make a movement towards the center of our
being and such a movement inevitably involves a
complete change of direction in the world. Through
practice we are led to relinquish the world so that
our innermost being may reveal itself. But once we
have experienced this innermost core and awoken to
our essential being, we begin to sense essence in all
things. And so, in the midst of our life in the world,
we become aware of Divine Being everywhere.

When essential being becomes inward knowing,
we feel a sudden change in ourselves. Relaxed and
free, full of strength and light, we are filled with a
new and creative life. Those things that lie heavily
upon us grow lighter; problems which have been
causing anxiety no longer oppress us; and events that
in our ordinary state would plunge us into despair
now lose their potency. Where every door seemed
closed all are now open. We who in ourselves were
poor now feel rich, and in the midst of clamor we
are peaceful and calm. It is as though we were bathed
in an invisible light which warms and shines upon
us. We sense about us a radiance that shimmers
through everything. But just as this radiance can
dawn upon us without warning, so can it as suddenly

disappear. We have no power to make it happen, nor to retain it once it is there. The most we can do is to learn to become prescient and aware of those attitudes which prepare us for such experiences and also, of course, those which prevent them.

The great lesson to be learned, when we are on the threshold of this process of transformation, is to recognize the vital importance of our inner experience, and to accept and admit to consciousness those moods and impulses through which Greater Being reveals itself. Transformation inescapably concerns not a part but the whole person. The time has come for us to comprehend the full significance of this statement. It means that a man must value himself— just as he does another person—in all his wholeness, depth and unity, in a word, as a subject. Only then will he be able to work rightly at the twofold task of cleansing his inner life and correcting his external attitudes. When we have understood that body and psyche are not two separate entities but merely two aspects of the whole way in which a person manifests himself, outwardly and inwardly, we come to understand that work on our inner being will inevitably effect the body, and that efforts to change the body will equally inevitably have their effect within.

*

Mention should be made here of two attitudes which particularly prevent self-realization. They are *hypertension* and *slackness*. The one reveals a person entirely in the grip of an ego which has completely lost touch with the powers that dwell in the depths of being and, because of this, is constantly concerned with its own security. The other discloses a person lacking in awareness and with a complete absence of any responsibility for that state which corresponds to essential being and is capable of revealing it.

Just as hypertension and slackness hinder all recovery from physical illness, so do they obstruct man's self-realization and inner well-being. We all come to recognize, sooner or later, to what extent the constant swing between these two conditions prevents the formation of the central core that is so essential to our being. But, once we are aware of the significance of such a realization, we cannot but strive to overcome these two great enemies of the right way of being.

Where there is hypertension, we find that excessive self-will, persistent self-control and constant surveillance by an all-too-watchful ego, block the flow of the forces of Divine Being which, in accordance with their own laws, inevitably pursue their course in concealment. In order to be able to operate freely, these forces require from man a basic trustfulness which enables him to give himself confidently to the life-giving and liberating Being which is present in his essence. All hypertension is an expression of a lack of trust in Divine Being.

Slackness, on the other hand, is a sign that man has let himself go and therefore forfeited his inherent form. A man can never be "correct" without a certain degree of awareness of the responsibility he bears for his total appearance (*Gestaltgewissen*). If he remains true to his conscience, it is impossible for him to be without form. The proper frame of mind will always reveal the combination of a trustful, passive attitude and an active attitude of individual responsibility. Most people have neither, therefore they lack the two most essential conditions for being in the right state of mind and body: *transparency and true form.*

*

Man, as a Person embracing body and soul, develops and realizes himself in every gesture he makes. But his personality is revealed most particularly through posture, tension and breathing. Nothing of what has been said, however, should be understood solely in a physical sense. Bad posture, wrong tension and incorrect breathing are all different ways in which a person's impermeability and lack of form make their appearance. Thus it is that we are able to assist into being this Person who, while embracing body and soul, is beyond the contradiction of these apparent opposites. Man can be helped to tackle the task of setting himself to rights by the proper practice of right posture, right tension and right breathing. It is the recognition of the need for this that opens up a field of possibility, accessible to all, through which the daily round itself becomes practice—practice that consists simply in learning how to be and how to behave in the world.

*

Man's correct posture is always determined by the right "center of gravity." The ingrained bad posture of many people is caused by the shifting of the center of gravity too far upwards. This can be seen in those who are dominated by the habit "Chest out—

belly in" where the drawing up of the shoulders is always an indication of tension. When we come across
such instances of incorrect posture, the reason is
never solely physical nor is it without importance.
Indeed, it has great significance, since it is a sign of
arrogance and an overemphasis on one's own person.
Such a posture is the defensive and protective mech-
anism of the man who is identified with his small ego
and who therefore seeks above all to safeguard himself and to hold secure his position in the world.
Whenever a wrong posture has become deeply ingrained it blocks the redeeming, renewing and
preserving forces that arise from the depths of Being.
The complementary form of this attitude, which is
often seen to alternate with it, is that of collapsing, or
giving way. In such a case, the innate image of man
as an upright being is lost. Such dissolution is evidence of a lack of feeling for, and responsibility towards, the correct form without which essential
being cannot manifest itself.

*

Once, however, we become aware of the harmful
effects of bad posture and consequently strive to develop and mature in accordance with our true nature
—which means an unflagging search for the right
center of gravity—we begin to perceive in ourselves
certain far-reaching changes. We know without
doubt that we are moving in the right direction.

The Japanese have a special term for that part of
the body which is the correct centre of man. It is
Hara. The word Hara, known to many through the
expression "Harakiri," literally means "belly." How-

ever, in its applied sense, it describes rather the general attitude of man, in which, freed from his small ego—released from pride, from the desire to dominate, the fear of pain, the longing for security—he becomes firmly anchored in his true center. Thus centered, he is able to be receptive to the forces of Being which, with or without his awareness, transform, support, shape and protect him while at the same time enabling him to reveal these forces to the world by the quality of his work, his ability to create and his capacity for love.

This effort to attain the correct center of gravity is the fundamental practice by means of which we are enabled to live in the world in the right way. Thus resting in the basic center, we are relaxed and free and at the same time feel ourselves supported. In the truest sense, here, we can be said to be upright. By means of this practice, the belly, the pelvis and the small of the back and their relation to each other form the basis of correct posture. When all movement flows from this relaxed, yet firm center, all gestures, attitudes and postures—walking, standing, sitting—become, as it were, testimonies to true being. There is no aspect of daily life that does not provide opportunity for this practice. If, for one moment, we forget Hara—whether it be in walking, standing or sitting—we cease at that moment to be fully and personally present. When the upper part of our body—the ego area—becomes tense and rigid, this rigidity cuts us off from our essence. On the other hand, when we sink downwards—dissolving into shapelessness—we lose the form which testifies to essential being.

All those actions which require concentrated

attention and will-power have the possibility of en-
dangering our hold on the center and our presence in
Hara. Whenever we aspire towards some goal or,
being dominated by our worldly-ego, we fix our attention on a particular objective, we are in danger of
losing touch with our center. Indeed, every kind of
work or activity which is aimed towards a definite
end can be the means—unless we are deeply
grounded in Hara—of shifting the centre of gravity
too far upwards. If we keep this in mind, it will be
seen that every action provides an opportunity to
practice correct posture. Each moment contains for
us the possibility of establishing and consolidating
the attitude which frees us from the domination of
our worldly-ego and allows us to exist from our essential being. To the extent to which we succeed in
this, whatever work we undertake will flow more
easily. Our knowledge and skill will be fully at our
disposal, our labors crowned with success and,
moreover, in relationships with people we will be
relaxed, uninhibited and free.

*

Tension and relaxation are two aspects of every
living whole. We moderns, however, tend to swing
backwards and forwards between the two mutually
exclusive extremes of these aspects—"Hypertension
and Slackness." Even when we talk of relaxation,
what we usually mean is nothing but complete dissolution—a condition which a moment later is inevitably transformed into a tension equally absolute.
What has to be learned and constantly practised is to
relax without being slack. Correct relaxation induces

the right kind of tension and has the effect of renewing us. The purpose of all true letting go is not some blissful condition where there is no tension at all, but rather a transformation by which a correct tension is achieved. It should be remembered that in this practice to achieve the right center, relaxation is never merely a physical or a technical exercise. It always implies a total transformation of the person. We need to learn to relax our *selves*—not only our bodies—in the right way. This entails far more than a relaxation of the muscles. Dropping the shoulders and letting go of one's *self* in the dropping of the shoulders are two fundamentally different movements. The one is a technical exercise, with results that are merely external and have no lasting effect; the other transforms the whole person. Hypertension bears witness to the fact that we are captive to the world-ego. Thus it is essential, in so far as we are identified with our world-ego, that we learn to let go not only at those times when we are tense, frightened, "up in the air" or defensively clinging to the upper part of our body —but at all times and in all places. This should be our primary concern.

*

Whenever we manage to drop that ego which holds us fast in a state of hypertension, it will be noticed that a change in our whole person is brought about. Normally, we become aware of wrong tension only when we are troubled by physical hypertension. But it should be realized that hypertension of any kind is always a sign of a wrong attitude to life. We have to learn to let *ourselves* go, as well as our attitudes. Only then, and only gradually, will we become calm and composed. It is impossible for us to relinquish, without falling apart, the support we gain from tension in the upper part of the body (the ego area), until we have learnt to anchor ourselves elsewhere—in a word, in the right center.

*

But that is not all. It should be remembered that the practice of right tension goes hand in hand with the practice of *breathing*. The inner significance of this practice of breathing will be missed if it is exclusively regarded as a bodily exercise whose sole purpose is to improve health and efficiency. When breathing means no more to us than the drawing in and letting out of air, it is an indication that we have not understood that, in truth, it is the breath of Greater Life which penetrates us and all living things. It moves through man, giving him life as a threefold entity—soul, spirit and body. By means of this breathing, man opens himself to Greater Life and closes himself again; gives himself away and again receives himself; loses himself and finds himself. When breathing is out of order, not merely the body but the whole person is out of order. Every disturbance in breathing signifies a disturbance of the total person and has the effect of barring man's way to himself. Wrong breathing impairs worldly efficiency. But what is more important, the deeply ingrained obstruction of essential being which it reveals affects the entire development. Continuous wrong breathing expresses a progressively strengthening inhibition of essential being. On the other hand, breathing that flows correctly indicates that the way is free for life to unfold out of essence.

The most widespread fault in breathing is found where a man breathes not from his center but from the upper part of the body so that the instinctive movement of the diaphragm is replaced by the activity of the chest muscles. In such a case an un-natural ego-breathing takes the place of that breathing which nature intended for us. This indicates that the man is consciously or unconsciously *doing* something which, if he were in harmony with his essential being, would happen automatically. Whenever incorrect breathing becomes habit, the process by which a man becomes a Person is obstructed. Such wrong attitudes to breathing are intimately related to other aspects of the personality, to all those fundamental misconceptions that man allows to master him and hinder his essential development in all spheres.

*

Throughout the centuries, man has invented many different kinds of breathing exercises that serve a variety of purposes. One may or may not agree on the question of their validity. But there is only one indisputable practice of breathing and that has neither been invented nor discovered by man. It is an essential part of him, inborn and in accordance with his essential being and consists simply in letting breathing happen! Such practice merely involves establishing—or rather, rediscovering—the natural manner of breathing. This means that the breath should be allowed to flow outwards and inwards naturally in a rhythmic movement of giving and receiving. With regard to the wrong habits we acquire through the domination of the ego—which lead us involuntarily to resist the full exhalation and then too quickly to draw in the breath—the basic and most helpful practice of breathing consists mainly in allowing complete exhalation. If this is done, correct inhalation follows quite freely without any assistance. Let me repeat that this practice—like that of relaxation—should not be approached solely from the point of view of a physical exercise. With the exhalation of the breath, we should let go of ourselves trustfully and without reserve. When a man can allow his breath fully to flow out, it is evidence of his deep trust in life. Whereas those who restrain their breathing mistrust both life and themselves! Thus— as it was with the practice of right posture and right relaxation—so it is with right breathing: it is above all a question of letting go the clinging, self-protecting ego. This is the only way in which we can free ourselves from our exaggerated inhibitions and hypertension and thus become receptive towards our essential being. In this union with being, it is possible for us to acquire the outer form which corresponds to our inner image.

*

Man's happiness as a person depends on the fulfilment of his deepest longings. In these he perceives Divine Being striving to manifest itself through him. His most fundamental yearning is for something which in his essential being he really *is* and is intended to be. His happiness, therefore, depends on the extent to which he is able to conform to his inner destiny. But such a consummation is only possible in so far as he freely fulfils this intention in his daily life.

The world in which we live is not a vale of sorrows which separates us from the peaks of the divine; rather it is a bridge which unites us with those peaks. We need but penetrate the obscuring mists that lie between us and consciousness and tear down the obstructing walls that bar our way. This necessitates living the ordinary day as practice. No special time need be set aside for this. Each moment is a summons calling us to recollect and prove ourselves. There is no activity, serving whatever external purpose, that does not contain an opportunity to dedicate ourselves more ardently to the search for truth. No matter what we are doing—walking, standing or sitting, writing, speaking or being silent, attacking something or defending ourselves, helping or serving others—whatever the task, it is possible to carry it out with a posture and an attitude that will more and more establish the contact with being. We must learn to make use of these opportunities in the practice of the "right gesture."

Healing Power and Gesture

In man the process of ripening and maturing, in contrast to the natural ripening of a fruit, is not accomplished automatically. It requires conscious collaboration. In the realm of instinctive drives, it is nature that thrusts us towards the fulfilment of our desires, and it is only necessary for us to yield to this irresistible force for such fulfilment to follow. The same does not hold good in the realm of the spirit; though here, too, the pre-condition for any change is that we are first seized by impulse. But unless man deliberately and consciously grasps that impulse and commits himself to achieving its realization, the current soon begins to weaken.

Spiritual progress inevitably brings about transformation. To this end it is necessary that a part of the former attitude to life be relinquished—some habit or requirement, some attitude or hard-won place on the ladder. In the long run it becomes indispensable for us to give up the prevailing convictions, apparently so natural, held by our world-ego. It is difficult for us to relinquish anything with which we have been habitually, though unconsciously, identified. Spiritual transformation, for this reason, needs a great deal of hard work. It not only entails inner practice, but also the practice of our manner of being physically present. This requires a long and patient prac-

tice of the body, of gesture, movement and posture.
One such basic practice whose purpose is transformation is that of meditation.

*

All genuine meditation consists of two stages. It
begins with concentration, by means of which a man
collects himself with the help of his will and directs
the power of his ego on to whatever may be the
theme and purpose of the practice of transformation.
Concentration is achieved through the powers of the
ego, which is able to grasp thoughts objectively, dif-
ferentiate between them and activate the will. Thus in
his concentration man comes to a closer understanding of the theme of his meditation. By means of this
effort he is able to recognize the faults in his physical
posture and the disturbing factors that are at work
within him. In addition it can bring comprehension
and a clear perception of the object chosen for meditation, no matter whether this be a picture, a word,
the following of the breath or some other theme.
The whole process takes place within the tension of
subject and object. Without such concentrated preliminary tension, without the attention that collects
the whole person—so that he is at the same time focused within himself and turned towards the object
—no meditation is possible. None of the foregoing,
however, is as yet meditation.

Meditation does not truly begin until that which
the ego had seized upon in turn seizes the ego and by
this movement changes rational analysis into synthe-
sis. By this means the quality of consciousness is
transformed from the masculine, active state which

is, as it were, a sword, into the feminine, passive condition of the open chalice that is capable of being
filled, as from an inner well-spring, without the active
help of the ego. Thus action is transformed into
"passion," and doing into being.

The word "meditation" comes from "meditari,"
which is a passive form of the verb meaning—"being
moved to the center," rather than the active which is
—"moving to the center." But this center is not a
fixed point on which a man concentrates as he would
on an object. Rather, the contrary can be said—that
it *itself* concentrates a man by drawing him inwards
and collecting him there. This center is, after all,
man's own essence—his transcendental core. Eventually, with practice, the center is experienced as a particular state in which the antithesis of subject and object is gradually dissolved. When this occurs the
meditator has the sensation of being centred. And so,
step by step, essential being awakes and becomes an
inward knowing that experiences itself as a personal
life centre anchored in Greater Life. This forms the
basis of a new consciousness of the world in which
man senses himself as being re-born.

*

The transformation which is the object of meditation is a progressive process. Identification with the
world-ego is superseded by identification with Divine
Being. But not until there is an integration of *world*-
ego and our *essential being* does the true self, i.e., the
Person, really exist. When this happens he is able to
reveal essential being in his changed, and now transparent world-ego. Thus meditation is not merely a

process of thought, but a transformation of the whole man; not only an inner happening, but one which includes the transformation of the body. It will be readily understood, therefore, that the theme of a man's meditation is less important than his manner of meditating. The prerequisite of any meditation worthy of the name is that the purpose of practice is understood to be progress along the Inner Way rather than worldly achievement or the enlargement of worldly capacities.

In meditation there is a close connection between what is done and how it is done. It may happen that the contents of experience have such overwhelming power that they literally throw a man to his knees, i.e., they force him into certain postures. On the other hand, there are postures of such permeability to transcendence, such power to transform that through them all contents of meditation become transparent to essence and Being. Provided a man is whole-heartedly committed in his striving for union with Transcendental Being, the releasing of the force which brings wholeness will depend more on the totality of his physical posture, than on the factual contents of his rational mind.

*

A man's state of mind at any given moment becomes apparent in his way of being present, his behavior and his gestures. Whatever posture he may adopt in the world, it will always be expressive of his total attitude. Man reveals himself through his gestures, but the same means that manifest him to the

world can also lead him to find self-realization as a
person. His gestures and general demeanor ring
true when they rise freely and directly from essential
being without having to pass through the barrier of
habitual and conditioned inhibitions. To the extent
that our gestures have been liberated from the prison
of our world-ego, the greater the quality of purity
they reveal. The greater the purity of the gesture,
the more we are helped to healing and wholeness.
For this reason it is possible to speak of the healing
power which lies in the pure gesture.

*

There is a story that tells of Master Eckhart's
meeting with a poor man: "You may be holy," says
Eckhart, "but what made you holy, brother?" And
the answer comes: "My sitting still, my elevated
thoughts and my union with God." It is useful for
our present theme to note that the practice of sitting
still is given pride of place.

In the middle ages people were well aware of the
inexhaustible power that arises simply from sitting
still. After that time, knowledge of the purifying
power of stillness and its practice was, in the West,
largely lost. The tradition of preparing man for the
breakthrough of transcendence by means of inner
quiet and motionless sitting has been preserved in the
East to the present day.[1] Even in cases where practice is apparently directed not to immobility but to-

---

[1] See Dürckheim, "The Japanese Cult of Tranquillity" and
Yasutani, "Zazen" in Dürckheim, *Wunderbare Katze*, O. W.
Barth.

wards activity—as in archery, sword fighting, wrestling, painting, flower arrangement—it is always the inner attitude of quiet and not the successful performance of the ways which is regarded as of fundamental importance.

Once a technique has been mastered, any inadequate performance is mirrored in wrong attitudes. The traditional knowledge of the fact that it is possible for a man to be inwardly cleansed solely through the practice of right posture has kept alive the significance of correct sitting. The inner quiet which arises when the body is motionless and in its best possible form can become the source of transcendental experience. By emptying ourselves of all those matters that normally occupy us we become receptive to Greater Being. As evidence here is a story of Dogen Zenshi.

The Zen practice of Dogen, founder of the Soto sect of Zen Buddhism, consisted in simply sitting still, without meditating on any theme or object. Esai Zenshi, the founder of Rinzai-Zen, on the other hand, mainly practiced the "Koan"—the solving of an insoluble mental problem. When asked for his opinion of the Rinzai method, Dogen replied: "It is very good." "But," protested the enquirer, "They practice the Koan!" "Well," said Master Dogen, "some people may be able to sit still only when they have something to think about. But if this brings them enlightenment, it is not due to their thinking but solely to their sitting still." The practice of keeping the body motionless transforms man's inner being. This story, like the one related of Master Eckhart, points towards something that needs most diligently to be explored.

*

It should be understood that the transformation which is brought about by means of meditation is not merely a change in man's inner life, but a renewal of his whole person. It is a mistake to imagine that enlightenment is no more than an experience which suddenly brings fresh inward understanding, as a brilliant physicist may have a sudden inspiration which throws new light on his work and causes a re-ordering of his whole system of thought. Such an experience leaves the person himself unchanged. True enlightenment has nothing to do with this kind of sudden insight. When it occurs, it has the effect of so fundamentally affecting and shaking the whole person that he himself, as well as his total physical existence in the world, is completely transformed.

*

To what extent the habit of sitting still can impress and change us becomes clear only when we have taken pains to practice it. After a short time we find ourselves asking: how is it possible that such a simple exercise can have such far-reaching effects on body and soul? Sitting still, we begin to realize, is not what we had imagined physical or spiritual practice to be. We are faced, therefore, with the question: "What is it we are really practicing if, although both are affected, it is neither body nor spirit?" The answer to this is that the person who practices, is himself *being practiced.* The one who is worked upon is the Person in his original totality, who is present beneath and beyond all possible differentiation into the many and

various physical, spiritual and mental aspects. In so
far as we regard and value ourselves as incarnate per-
sons, certain manifestations in our life move from
their accustomed shadow into the light of understanding. Thus our moods and postures take on new
meaning. So long as we think of body and soul as
two separate entities, we regard moods simply as
"feelings," and look upon bodily attitudes and breathing as merely physical manifestations. When, however, the whole person is recognized as a "thou," it is
no longer possible to separate the body from the soul.
Once it becomes a question of transformation, our
basic inner moods, together with all the gestures and
postures that express them, acquire new significance.
They are the means through which we grow aware
of, manifest ourselves, and become physically present
in the world.

*

We have seen that the purpose of all practice is
transformation. It is the means whereby essential
being is able unequivocally to prevail in man's inner
and outer life. His new state of mind and the radiance that now floods his ordinary existence can permit Divine Being to be ever more clearly manifest—
not only in his inner impulses and his newly-
achieved sense of purpose but in the sense of
blessedness that pervades everything he does. In his
essential being man participates in Greater Being,
and since it is this participation which gives life to his
own being, man's transformation brings about the
manifestation, in the world, of the Transcendental

Being present within him, amid the conditions of his ordinary life.

*

During the process of man's growth from child to adult, he acquires the kind of consciousness—with its corresponding attitudes—that in effect cut him off from participation in Divine Being. He develops an ego capable of the kind of logical thinking that enables him to confront and rationally master the world. The world-ego, as this "I" is called, views life solely as a matter of establishing and sustaining certain fixed positions. It follows that when man is orientated theoretically, practically and ethically solely towards established facts, essential being is lost. Life, being a moving process, does not tolerate anything static. The loss of contact with Greater Life becomes evident in man's relationship to the world, wherein he lopsidedly develops only those of his abilities which help him rationally to understand and dominate his surroundings. Such obstruction of essential being culminates in the delusion that he is himself autonomous, able to control life and give it meaning solely through his own powers. This delusion creates a barrier which separates man from Divine Being plunges him into a specifically human suffering. Yet it should not be forgotten that this same suffering, which is the result of estrangement from Divine Being, can itself become the means whereby Divine Being may at some time shine out against the background of our misconceptions. For, when the suffering caused by estrangement has grown so great that man is overwhelmed by despair and by the fear of being cut off from his roots, he may, at that moment, find himself ready to open to Divine Being which is all the time striving to deflect him from his lopsided concern with the world, and turn him towards itself.

*

The conditions necessary to make possible this reversal in a life that has been orientated solely to the world, are: (1) that the reality of Divine Being present within a man's being should ceaselessly strive to become manifest in and through him; (2) that this man with his rationally orientated mind should have become estranged from and be in conflict with his essential being. Without this tension, this tug of war, man would be as little able as a flower to become aware of the Divine Ground of his being; (3) that in breaking through the barriers of what is inevitably a limited awareness, he opens himself to Greater Being and thereby awakens to the realization that the barrier has its origin solely in his rational ego; (4) that he is prepared to allow the Divine Will which summons him from within, to become his own will, and to *decide* to follow the Way to which essential being calls him. This reversal of direction—arising as it does from his suffering at being estranged from essential being and his joy at rediscovering it, signifies his setting out upon the Way.

*

The Way is the never-ending practice which leads us from the reality that was shaped by our world-ego

to the reality that is beyond time and space, and
thence towards transparence and new becoming.
Only the man who, with the full force of his nature,
realizes that the whole meaning of his worldly existence lies in the manifestation of Transcendent
Being, will be able to reach the Way. When he understands this he will need to decide once and for all
to become a witness to Divine Being in his ordinary
life, to whatever extent it is given him to be so. He is
thus no longer bound in vassalage to his world-ego,
nor even, in the final instance, to the advancement of
the world and his fellow men. Beyond all these, he is
committed to the service of a new master—Divine
Being. In so far as this new master is rooted in inner
experience and is not merely the result of blind belief, this service presupposes perpetual contact with
essential being.

*

We may be moved by Greater Being in many
ways and in different degrees. A yearning, of which
we are almost unaware, may perhaps stir in our
hearts. It is not only the great experiences of "breaking through" that completely transform our relationship to life. The smallest incident may equally
achieve it. The primary aim of all practice on the
Way is to prepare man for such experiences of Being
and by making him receptive to genuine contact
with essential being, open for him the Way of Transformation. The purpose of this practice is to help
man to bring to an end his separation from essential
being, and lead him back to the path of reintegration with his own inner being. This process

serves man's greatest well-being (*Heil-Werden*). But what do we mean by well-being or, in other words being whole?

A man may be thoroughly efficient and skilful, able to establish contact with and adapt to the world and thus successfully master his life, but if he is still deaf to his essential being, it does not yet mean that he is well and whole. He can be said to be whole only in so far as he is open to the redemptive Ground of Being, and at the same time able to accept and obey the forces of renewal that give him direction and freedom. The man who, in the midst of all the confusion, poverty, and disunity of the world, is able to manifest the plenitude, significance and oneness of Transcendental Being is indeed whole. There is no one, however, who can finally achieve this state. All any man can do is set himself on the path toward it. He can be said to be healed (or whole) once his contact with Being has produced a mode of life which keeps him on the Way of Transformation. He who practices ceaselessly so that all gestures, postures and movements (which is to say, his bearing and ability to yield himself up) are a true expression of his essential being—that man is healed and will be enabled to remain on the path that leads to wholeness. Thus, the purpose of correct practice is not to bring man to a state of tranquillity but to keep him in a condition of constant watchfulness and prevent him from coming to a standstill on the Way.

*

The so-called "peace" of the world-ego, illustrated by the bourgeois aim of a "quiet life," comes about

when all inner movement and growth have stopped.
Of quite a different quality is the peace of inner
being and the life which strives to manifest itself
through it. This kind of peace can only prevail
where nothing further interrupts the movement towards becoming. To achieve such an attitude to life
is the aim of all practice and meditation; it can never
represent a state of "having arrived" but is always a
process of "being on the way." Such practice, therefore, is by no means acceptable to all. There are
many who throng to the so-called prophets who
promise a cheap kind of peace to troubled mankind.
But such "masters" simply betray man by hiding
from him the real cause of his anxiety, which lies in
the desire for transformation inherent in his inmost
being.

*

How can we recognize the correct state of mind?
What are the obstacles that stand in the way of
achieving it? Strangely enough, the very manner in
which a man develops his I-world awareness is one of
the chief obstructions. This is a form of consciousness in which the ego, intent on asserting itself, con-
solidates its rational view of and position in the
world. Thus we see that the danger with which such
an ego threatens the maturing man lies not only in
the well-known threefold sin of seeking possessions,
assuming importance and striving for power! The
real cause lies rather in the fatal human tendency to
remain static. The so-called natural ego is ruled by its
inclination towards and capacity for holding fast to
fixed positions that seem to promise security.

Thought and energy are turned towards facts that have been rationally established, whether their purpose be to safeguard possessions in the world, to defend a system of thought or to cling to fixed values. Whenever the idea of holding fast dominates the mind, life itself is in danger. But where life is, nothing is fixed, all is in motion and engaged in the eternal process of crystallization and dissolution, growth and maturity, death and rebirth. If life is present, then even that which appears to be completed is open to further change, and that which may have finished growing still remains transparent and ready for new becoming. So it is with the man who has awakened to life. Even those attitudes which have been established in him by the rational mind are permeated by the constant movement and transforming power that come from essential being.

Whenever essential being makes itself felt, an attitude is created in which the whole static order of the world-ego is taken over by a new and dynamic process of continual becoming.

*

What is it that we mean when we use the phrase Divine Being? Divine Being is an abstract and, as it were, far away term in relation to our conceptual thinking. But in the depths of our inner experience it is the one most concrete and closest to us. It is, after all, that with which in our essential being we are one; or, to put it more clearly, it is that which in our being we really *are*. For this reason it is able to shower upon us experiences of such force, radiance and plenitude, that our rational mind is completely overwhelmed.

When we are thus seized by Divine Being, our basic mood entirely changes. Strength, joy and love seem to wake within us—experiences which, from the standpoint of the ego, are incomprehensible and without reason. Inevitably, the impoverishment, senselessness and uncertainty of our ego-centred worldly existence show up sharply and painfully when seen against the background of our first experience of Being. Henceforth, the responsibility for revealing Divine Being in all we do weighs even more heavily on us than does our former desire to enlarge ourselves in the world. Even so, whenever Greater Being stirs within us, through all the despair we experience when we recognise our imperfect state, and all the pressure of our newly-awakened conscience, we feel a radiant, warm and creative purpose streaming through us. But if Greater Being is to be enabled not merely to move us for a brief moment, but constantly to be made manifest *through* us, it will be necessary for us to find a way of life that extends beyond the moment—a way that makes us new in body and soul. This we can only discover and establish with the aid of ceaseless practice.

The experience of Divine Being and the transformation that is the result of contact with Divine Being are two different things. In the matter of the transformation that comes from Divine Being, we mean more than the mere acquisition of a new content to life or a new scale of values—more even than the step that leads from rational understanding to experiences

rich in inner images. It is possible for a man to have
wonderful visions that herald the advent of Divine
Being, without really tasting it. He may experience a
foretaste of Divine Being in a dream or in a moment
of great happiness, but such momentary influences
will not be sufficient to transform him. In fact, he
may have countless revelatory experiences without
being changed by them one whit. They may even
happen daily and leave him essentially unaltered. It is
only when we really hear the summons that resounds
in all genuine experience of Being and, having heard,
commit ourselves to reply, that our state of mind becomes sufficiently modified to correspond to it. In
this way the new Person is born and life is filled with
fresh meaning and impulse. Imagination, action and
thought are now so impregnated with the transcendental Ground of Being, that even the authority of
the rational perception undergoes a fundamental
change. It is no longer the ultimate judge of our
understanding and the arbiter of all our actions. We
must not, however, because of this, revile and con-
demn it. The rational mind, after all, is that which
distinguishes us from the beasts and is the prerequisite for spiritual growth. Neither rational thinking,
nor its focal point the ego, are in themselves evil:
they only become so, when man identifies himself
with them alone and thus becomes estranged from
the Ground of Being which is by its nature beyond
rationalization. The real significance and purpose of
all rational behavior should be to make room for
Divine Being to unfold within us and, through us, to
manifest itself in the world!

*

To change our attitude by means of the practice that accords with essential being requires, above all, that the world-ego be dislodged from its central position. Once we have realized this, it necessarily follows that in all those areas of life and experience which cannot be subjugated by the grasping, fixing ego, our interest is enlarged. Two realms of experience now acquire particular significance—that of sense and that of inner bodily responses. Whenever our hierarchy of values is ordered by rational principles these take a back seat. But they can never be robbed of their quality of immediacy and freshness, even when they are experienced within the frame of a rationally ordered life. Their intrinsic quality enables them to become in us beneficent witnesses to Divine Being, provided we learn to live with them and let them be. It is a truism that man, even when enmeshed in his rational activities, is continually renewed by his contact with nature—country air, forests, the sea shore. Practicing the perception of the qualities of smell, taste, sight, hearing and inner bodily movement can have a wonderfully refreshing and curative effect so long as we are able to experience them purely. By "purely" we mean free from the ego's habitual process of labelling and naming all things according to their specific properties. There is a form of meditation in which, by sharpening our senses, we unlock the transcendental content that resides in them and thereby assist transformation. Even more important, however, than this freeing of the senses, is our need to experience our bodies—always provided that the body is not regarded merely as a thing, but as the means by which a man manifests himself in the world as a Person.

*

Whether it be in repose or in motion, the human body is intended to carry, transmit and bear witness to essential life which is designed to assume a particular form in the world in accordance with its inner image. In this way, the body bears witness to that Being which is ever striving within it towards manifestation. The word "body" here is not to be thought of in static terms: it implies rather the unity inherent in the changing pattern of posture, attitude and gesture by means of which an individual person lives out his life. This body, however, since it is subject to the distorting conditions of the world, can never become the perfect expression of Being, nor the complete realization of that inherent image which is unconditioned by time and space. Whatever its momentary form may be, it will always mirror the imperfections that the world has wrought in it. A man's physical state and the gestures that reveal it can give, on the one hand, clear indications of the measure and manner of his accord with essential being and the extent to which he has become himself; or, on the other, it can provide indubitable evidence of how far he has failed himself.

*

In looking at man simply as man we should not make any distinction between his body and his soul. If we are able to set aside this inevitable antithesis, we see that his most direct revelation of himself is through his gesture. Indeed, his whole field of move-

ment, posture and facial expression provides incon-
trovertible evidence of his inner and outer states. In
respect to our work when we are on the Way, the
manner in which our bodies move is very significant,
for it is here that we experience ourselves not outwardly but inwardly. To observe and register what
happens from a purely external standpoint is very
different from inwardly experiencing the rhythms of
the breath, and the inner sensation of movement.

*

Rational knowledge has not the same quality as
inward knowing. On the other hand, if we are to
achieve clarity in our inner awareness, a background
of acquired knowledge is essential. The importance,
for the practice of the Way, of the inward knowing
of the body's changing processes, can only be realized when we understand that this body, in its existential sense, differs from the body comprehended
analytically, i.e., as an opposition of spiritual and
physical forces. The bodily existence of a living man
has no relation to the corpse he leaves behind when
he dies. In its own way the body's manner of existence shows how far a man's transcendental being has
been able to express itself under the conditions of
space and time. The common view, which differen-
tiates between body and soul, is the product of the
analytical, rational mind that cannot, by its very nature, perceive the unity of the living person.
Whenever man is looked upon as a thing—a thing,
moreover, logically determined—we lost sight of him
as a Person. It follows that if we do not perceive him
as a unique, living whole, we not only cannot help

him on his way to self-becoming, but we fall into the
fatal dilemma of dualism. With its rational distinction
between the physical and spiritual aspects, dualism
separates man into two different entities. This logical
view can so mislead us in our practical evaluation of
people that we tend to regard another person as important only in respect to the significance of his
functioning in the world and to ignore his existential
requirements. To reverse such methods of thinking is
far more difficult than is generally supposed. High-
sounding platonic assumptions of the unity of body
and soul are no more useful here than our ineffective
modern attempts to co-ordinate the practices of medicine and psychology. We shall not achieve any valid
understanding of a human being by adding what we
can deduce from the living corpse to our knowledge
of the bodiless psyche. The concept of the unity of
body and soul is the product of discursive thinking.
It fails to take into account that vital Person who is to
be met in another as a *thou* or experienced within
ourselves as an *I*. Such an attitude makes it impossible, either theoretically or practically, to come to
any understanding that could do a man justice as a
living Person.[2]

*

When we meet him as a Thou, when we are able
to see him in his existential relationship to the world
and to life—only then does man as Person appear. It
is possible now to see him as an element in a system

[2] See Hans Trüb, *Heilung aus der Begegnung* (Healing
through Encounter), Kluett-Verlag.

of interrelated parts whose order has no relevance to
the order so dear to the rational mind. We can, indeed, never do a man justice with our rational understanding alone: our direct participation in his life and
suffering is necessary for this. Only when he is present to us as a Thou can we feel and sense him in his
personal, human existence, where he, like ourselves,
is striving for happiness, meaning and fulfilment.
When this happens we recognize in him a brother
who is also on the Way, borne along, by destiny, as
we are, towards the goal of becoming a true man—
which is to say, a Person. A doctor for instance, will
lose sight of the human being if he looks at the patient simply as a case.

*

At every moment in his life, man as a Person is
bound by his own essential law to pursue his transformation according to his inner destiny. It is in the
context of this law that his particular bodily form is
to be understood and evaluated. By this means it is
possible to discover whether he has strayed from or
remained faithful to the Way of Transformation;
whether he has moved forward, or has become stuck
fast at some inhibiting point. Any incorrect bodily
form, any hypertension, any cramped condition—
understood existentially—is an expression of an ingrained wrong attitude. It is an indication that a man,
as Person, has gone astray or been brought to a standstill on the Way.

*

The dominating tension, existentially speaking, is that which occurs between man's world-ego and his inner being. This corresponds to the tension between his bodily form which has developed under the circumstances of space and time, and his essential form, as yet unrealized, but which is nevertheless always absolute in its demand for realization. The correct attitude necessary for the fulfilment of his life as a Person is one in which his conditioned body has become transparent—in other words, made permeable for the revelation of his essential being.

*

That which, above all else, stands in the way of the realization of the person, is the fixed attitude—or crystallization—by means of which man attempts to establish his security and to assert himself in the world. The physical expression of such a fixation is easily apparent in anyone whose ego is striving to attain security. As we have already said, it is revealed particularly in the posture, in the relationship between tension and relaxation and in breathing.

When we speak of posture, tension and breathing, it is necessary to make a distinction between their significance as man's personal expression and their value as data for the rational-analytical view—as, for instance, in the practice of medicine. When we consider man existentially we accept his demeanor and appearance as ways in which he manifests himself. We understand that it is by these means he is able to be himself and also to be present as himself in the world; whereas, regarded from a logical or rational standpoint we may be aware only of some deformation of the body or of inner obstacles that inhibit his strength and efficiency. If our concern is with self-becoming we cannot fail—for the sake of our own understanding—to take into account the personal significance of posture, breathing and patterns of tension. No exercitium can ignore these three elements, for the reason that man's physical attitude does not represent only a bodily state, but also his way of being present as a Person. And there are, as we shall see, exercises which serve the inner development of the Person by bringing about the correct bodily form.

The manner in which he is physically present and the degree of his permeability reveal the point at which a man has arrived on his way to maturity—which is to say towards integration with his essential being. He may perhaps be physically ill and yet, as a Person, present in the right way because of his permeability to the demands of his essential being. In contrast to this, a thoroughly healthy athlete may well be present in the wrong way if his inward growth is obstructed by arrogance or self-love, either of which will bar him from essential being.

*

A man's correct attitude is never static; it is not something that is achieved once and for all. Rather, it is a living, moving and changing process within which he can remain receptive towards his essence, in spite of the fact that the bodily form he may have at any given moment is bound by conditions of space and time. This enables him to obey the striving of his inner being to become manifest, to remain always in a state of transformation and to become ever more permeable to Divine Being.

An attitude, or state of being, manifests itself and develops from moment to moment in the unity of a man's gestures. As we speak of correct attitude, so can we speak of transparent gesture—which is one capable of manifesting the right attitude and, by repetition, becoming firmly established.

This transparent gesture may be said to exist when essential being is fully manifested in it. To make this possible, the gesture must be free of all fixations grounded in the ego, and above all from the dominance of those attitudes by means of which the world-ego protects itself. Our concern is to become transparent to Divine Being and to let it resound through us. The pure gesture can manifest essential being in accordance with the inner image. But no gesture, even though it may have in it nothing of ego, is really pure, so long as it lacks form. It is by means of the pure gesture that the correct form is realized and able to prove itself. Such a gesture is not inspired by the ego, but is a gift to man from his inherent image. Thus the correct physical form is the means by which the image can express and fulfil itself.

*

By the purity if his gesture we are enabled to see that a man is present in the right way. This means that he is permeable to his individual essence as it strives towards the revelation of plenitude, innate form and unity of Divine Being. When a man is present in the purity of his gestures, his chief characteristics are a simple, unaffected trustfulness, a natural harmony, and an attitude of friendliness which can only exist when there is genuine contact with essential being. In a like sense, the pure gesture also liberates the healing spirit of which it is an expression. It is both the manifestation and the ever-living source of an attitude which enables man to respond appropriately to all situations of life. Each new circumstance stimulates the cycle of transformation afresh. Thus we see that the pure gesture is an expression of the law of movement which is constant renewal. By means of pure gesture, man as an individual "form in time" (*Zeitgestalt*) is enabled to relate in his own way to the primal movement of life, wherein the forms which have crystallized in him under world conditions are repeatedly dissolved and remade in accordance with essential being. This is a process of continuous transformation.

The pure gesture is evidence of the fact that man has found a vital way of life, a way which enables him to face the world fearlessly because he is living directly from essential being. Thus, the more unequivocally a man lives from essential being and without regard to his world-ego, the more freely and without prejudice will he meet all that may, from moment to moment, confront him. This means that he is really open to life which, as we know, never repeats itself. Now he perceives even familiar things as if for the first time; no longer taking up fixed positions, he is able to give his whole self, just as he is, to every passing moment.

The practice of pure gesture requires first that a man have a continuous awareness of his usual behavior, his prejudices, habits, and processes of self-protection; and secondly that he should awake in himself a determination to retract and dissolve all these attitudes with which he deceives himself as to what is the truth of life.

The Wheel of Transformation

It is impossible for man to achieve self-realization as a person, unless he is prepared fully to co-operate in the process. At each step this process depends on man's whole-hearted concurrence, for in the long run self realization can only be the result of his constant and tireless practice. Not until a man has begun to practise continuously with complete awareness can he be said really to have joined the Way. From then on the Wheel of Transformation never stops turning. Or, if it does stop, it topples over and man becomes stuck fast and is unable to move.

· The process of transformation consists of three stages:

1. All that is contrary to essential being must be relinquished.
2. That which has been relinquished must be dissolved in the Ground of Being which absorbs, redeems, transforms and recreates.
3. The newly formed core which arises out of the Ground of Being must be recognized, accepted, allowed to grow and personal responsibility for it undertaken.

Further—since that which conflicts with essential being must first be discerned in order that it may be relinquished—so the newly-formed core must be assented to by the will, in order that it may ultimately be realized and made effectual in life (reintegration with the world-ego).

To this end, it may be said that the Wheel of Transformation has five spokes (steps):

1. The practice of critical awareness.
2. The letting go of all that stands in the way of new-becoming.
3. Union with the Ground of Being.
4. New-becoming in accordance with the inner image which has arisen from the Ground of Being.
5. Putting to the test this new form, practicing it in everyday life and noting all failures by means of critical awareness. And so the wheel turns.

Each section of the revolving wheel serves in its own way to create an attitude that enables man to grow more and more transparent to Divine Being. To achieve this, no step may be omitted. Each one, if it is truly accomplished, contains all the others. Yet each has its own specific purpose. Here it should be said that it is helpful for the beginner—and which of us does not remain a beginner throughout his life?—to lay special emphasis, in his daily life or in set periods of practice, sometimes on one step, sometimes on another. The essential thing to remember, however, is that transformation is only possible so long as the Wheel is kept turning. The steps have their use and meaning only within the context of the continuously revolving wheel.

One isolated experience of Divine Being, when in some unique moment we are possessed by feelings of liberation and responsibility, is, of itself, never enough to give us the strength to keep the wheel in motion. It is essential that we recognize and repeatedly grasp anew the proper purpose of such experiences—which is to bear witness to Divine Being in the midst of our life in the world. Further, we must strive for the fulfilment of this purpose by repeatedly affirming our decision, rooting it firmly in our conscience and absorbing it into our will. Only from fidelity to this process of ceaseless transformation can critical awareness, letting go, union, new-becoming and our testimony to these in daily life, be existentially valid.

1. Critical Awareness

Critical awareness bears upon all that obstructs our contact and integration with essential being, as well as all that stands in the way of true self-realization. Critical awareness is not a state of mind which can be reached once and for all, without further effort: it is a continuing process which has as its aim the growth of consciousness. It is not to be confused with the purely intellectual faculty but is, rather, an inner awareness, an inner knowing—indeed, a transforming power.

For instance, to take cognizance intellectually of the fact that there is a contradiction in what we think of as the ideal image is not at all the same thing as becoming inwardly aware of a wrong attitude. The latter process is, rather, an inward sensing of the way in which, as a person, we are rightly or wrongly present—that is, in harmony with or antagonistic to our true nature. This sensing grows more accute the more completely we are committed to following the Way, to living in accordance with essential being and to obeying our inner law.

The creative and transforming power of critical awareness develops out of an inner awareness of the body, and not from some rational conception of what we believe corresponds to or conflicts with correct posture. For this reason transformation begins only after the necessary attitude has been inwardly experienced. Thus we find that, in order to be effective, our orientation towards the Divine and its manifestation—which should be the foundation of all other actions—requires far more than the conceiving of an ideal image and systematically striving towards it. It is not until we comprehend our inner structure that such orientation effectually begins to permeate all bodily movements. Only by means of continuous single-minded effort can this process in the long run either bring about transformation or exclude all that stands in the way of such permeability.

*

With regard to the development of critical self-awareness, what element in it should we consider most important to practice? It is the awakening of, the separating out and, as it were, the making articulate a new sense that can smell out inner truth. Such work requires the development of a sensitivity which will enable us reliably to perceive all deviations from the true inner order. In particular we need to become susceptible to that within us which knows beyond any doubt whether we are wrongly or rightly centered. In this way critical awareness, being the means by which we progress, becomes active in each aim and aspect of a man's life. Through it we come to understand that what is required is the development of a new kind of sensitivity which will enable us to sense the correct center and to become aware of all deviations from it.

*

Few people, in these days, have any feeling for or understanding of the right center, even though there may be considerable technical knowledge available about right and wrong posture. Being correctly centered does not mean merely that a man is so anchored in some central part of himself that he can, from that solid position, confidently confront the world. This is not enough. The right center is that which enables us to correspond to essential being. Once a man has been truly seized by essential being and undergone the transformation that results from such an experience, his life will henceforth revolve around his determination to attain, preserve or re-establish that state of mind and body which can remain permeable to essential being. This revolution around the true center begins only when man is able to sustain the proper inward attitude without slackening. All distortion and loss of the correct physical form—that form through which man is intended to fulfil his life—is connected with the absence of the right center. But what do we mean by the term "center?"

In relation to the "what" consciousness, which sees reality solely in the context of "things," the word "center" has merely a spatial or physical meaning.

For the "who" consciousness, however, "being centered" signifies a state wherein a man moves continuously toward his innermost nature; that is to say, his aim is centered in his essential mission which is to manifest Being in the world. The opposite of such an attitude would be one where a man is exclusively orientated toward worldly values and whose life revolves wholly round the desires of his world-ego.

In other words, "center," from a rational point of view, refers to a fixed point. In its personal meaning, however, the center is the center of our individual life, our driving force, reason and purpose. Understood in this personal sense, the word "center" has a threefold foundation. It stands on the *strength* by which we live, the *meaning* around which we revolve and the *fulfilment* for which we search.

*

Inevitably man's life is focused and nourished by a specific vital force, by search for individual meaning and by a longing for personal fulfilment. Strength, meaning and fulfilment, however, have a different significance according to whether a man identifies himself with his world-ego or with his essential being. Strength from the standpoint of the world-ego is a quality which enables man to survive healthily and efficiently in the world; meaning may grow out of any kind of work—including service to the community—that is valid in and recognized by the world; fulfilment is brought about by whatever finally sets the mind at rest. When, however, man is identified with his essential being, strength is the result of contact with essential being, meaning derives from obedience to the requirements of essential being and, once he joins the Way of Becoming Transparent, fulfilment is at hand. In a state of identification with his ego man is dependent on the world; he needs the world and relies upon it to give himself meaning. The development and functioning of his worldly strength, as well as his self-confidence and ability to survive, are conditioned on all sides by external factors. Under such conditions he is liable to be easily upset, touchy, quick to feel aggrieved and always on the look-out for security. He sways backwards and forwards between states of restriction and abandonment, hypertension and dissolution, elevation and prostration. He experiences every condition except that of being firmly balanced. Why is this? The explanation is to be found is his lack of a center capable of supporting him, a center which cannot be shaken. Once a man is anchored in his essential being, he becomes aware that there exists in him a core that nothing can destroy. From this he gains stability and permanence. He acquires a composure that is independent of the world, a clear sense of his inner direction, and above all, a self-confidence that is independent of the world's praise or blame. The personal significance of "being centered" is that a man can so live in the midst of all the ups and downs of life that he receives strength, purpose and direction from essential being. Imperturbable and at peace, he ceaselessly pursues his inner destiny and so manifests Divine Being in his life in the world.

*

Stretched as he is between heaven and earth, man's spiritual development and the possibility that essential being will reveal itself to him depending, in the first instance, on his being correctly anchored in the "earth." In other words, he has first to discover his right center of gravity. Understood in the personal sense, "center of gravity" does not refer to any specific area, but to what we may call the root of personal or existential life. Nevertheless it is possible to perceive whether or not a person is properly rooted in his physical body. When a man has found this correct center of gravity in his body he can open himself to the forces that lie at the essential core of life and anchor himself therein. The "Vital or Earth Centre of Man" [1] contains all the energies of the Ground of Being, energies that bring to his aid a sense of liberation, a strong inner support and the possibility of regeneration. When he is relaxed and firmly based in this center man experiences these forces in his physical existence. He immediately loses them, however, when he draws himself up tensely, for such a movement cuts him off from the life-giving strength from below. The center of gravity is thus misplaced. Worse still, it can happen that no center of gravity exists at all—except perhaps for a fleeting moment—in which case hypertension or dissolution are the result. It is the center of gravity which determines the total posture of the body as well as its patterns of tension and the character of its breathing.

*

[1] See Dürckheim, *Hara, the Vital Centre of Man*, Allen and Unwin.

That power which enables us to be truly centered lies, physically, in the middle of the body, in "Hara," or more accurately, in the pelvic region.[2] Hara (Japanese for belly) refers to an attitude by means of which man is anchored "below" in such a way that he is freed from habitual restrictions brought about by being top-heavily centered "above" in his world-ego. This settling-down into himself "below," this re-orientation of body and soul, allow the strength within man's own being to support and mould him, and to give direction to his life.

A man may have an incorrect center of gravity or even none at all, and in either case this is a result of the extent and dimension of his world-ego—which is to say that he has either too much of this or too little. The more he identifies with this ego, the higher up in his body will be his point of anchorage. Caught thus in these extremes of the ego, his attitude will be determined by his intellect (head), his will (chest) or his all too sensitive feelings (heart). Thus drawn up, tense and rigid, he will be totally at war with his instinctive drives and cut off from cosmic forces. If, on the other hand, a man's ego is underdeveloped, the connection to the center from which strength and form are derived may be almost completely lacking. As a result, he will fall apart of collapse, and become the plaything of inner and outer forces. He will be unable to be true to himself or to stand up to the world as a person. Thus we see that when the center of gravity has become fixed too far upwards

[2] There is a French expression: "Il n'est pas dans son assiete." Assiete, "plate," means the pelvis. When one says of someone, "He is not in his pelvis," it signifies that at present he is hypersensitive, irritable, unrelated and easily disturbed. In short he is unbalanced etc.

in the body, man is cut off from the Ground of
Being as well as from the liberating and supporting
power of his own essential being, and thus is trapped
in all the menace and terror of the world. When, on
the other hand, he has no center of gravity at all,
even a minimum amount of selfpossession will be
lacking; he will have no means of receiving the in-
coming stream of Divine Being since he has not developed that inherent form through which he could
give it entrance and pour it out again in his work in
the world.

*

The core of the practice of pure gesture lies in
learning to anchor the whole of ourselves in the correct center of gravity. It is vital to awaken within
ourselves an innate feeling for the true earth center.
This should be our constant practice so that, with the
help of critical awareness, we may bear witness to it
in our lives.

No matter how manifold nor how complex the
manifestations of the earth center may be, it never
ceases to summon us and entreat. If we are prepared
to listen faithfully, we will constantly hear its call. It
rings out with equal felicity and promise, no matter
what the situation: in conditions of joy or grief,
dearth or excess, and in the warning voice of conscience that cries aloud to us when we lose sight of
our inner destiny.

*

As we have already seen, the counterpart of a
wrong or a non-existent centre of gravity is a tend-

ency towards hypertension, with its inevitable compensation, dissolution. Hypertension is an indication that the attention is predominantly concerned with seeking to consolidate a particular position. This kind of fixed attention will never relax its hold while man is held fast by his ego. Every type of hypertension, every kind of contraction, is an expression of lack of trust and reveals, whether actively or passively, attitudes of self-protection or resistance. Such attitudes cannot, therefore, be corrected by purely technical means: only long practice of the fundamental gesture of trustfulness serves any purpose here.

Hypertension always goes hand in hand, as it were, with the kind of shallow breathing that takes place in the upper part of the body. This of itself is sufficient evidence of a man's imprisonment in the ego. It lacks the calmness and composure which distinguishes one who, being rightly centered, is in direct contact with Being. Here, critical awareness proves its value with its capacity for sensing the wrong center of gravity, incorrect tension and inharmonious breathing. It is important that this sensing of the center of gravity, tension and breathing should not be thought of (or practised) in any rational way, but understood rather as an inward and personal attitude. Only thus can it serve as a means of fulfilling man's basic, inner law, which requires him to be permeable to Divine Being and to live from his essence in his own particular form. But the ability to sense what is incorrect, as well as to establish what is correct, will only be present if the larger, existential search has already begun.

*

As long as a man, although able theoretically to recognize what would be for him the correct state of being, still thinks in terms of the desires and anxieties of his ego—seeking freedom, comfort or security with his rational understanding—he will not be able to perceive the personal significance either of the center of gravity, or of tension and breathing.

Under such conditions, he will reason solely in physical terms. He will tackle wrong posture in a purely technical way, attempting to correct it simply as a pragmatic necessity, for the sake of his health and efficiency. In addition, he will mistake the practice of establishing the right center, and the firm anchoring in the pelvic region, for some physical trick that will be of use to his world-ego. In such a case, the practice of the right center, of stillness and of relaxation—if carried out merely for the sake of easing the mind—will turn into the opposite of what was intended. It will become the means for enabling man temporarily to retain his wrong postures and attitudes without discomfort, and thus to remain caught in the ego, which by its nature unfailingly tries to turn its back on suffering. The practice that relates to Being is quite different. It makes it possible for man to risk himself courageously, frees him from his fear of the world and, through his acceptance and endurance of the harshest tests, enables him to remain on the Way of Transformation. On the other hand, wrong practice will always keep him at a distance from Divine Being.

2. Letting Go

The more sensitive we are to those incorrect inner attitudes that signify an equally incorrect approach to life, the more it is a sign that we are interpenetrated by the Divine Presence. It can even be said that this presence sets free in us a harmonious and receptive form. A new feeling of conscience towards our way of being emerges and this has as its corollary an increase in critical awareness. Eventually, this sensing of what is erroneous in our attitudes coincides with the impulse to let go that which is wrong or stands in the way of the right attitude. The action of steering a bicycle or car provides an illustration of this. The continual instinctive adjustments that are necessary if the vehicle is to be kept on a straight course cannot be separated—any more than cause and effect—from the driver's awareness of every minute deviation from the chosen direction. The more practiced he is, the more does his recognition of such deviation coincide with his adjustments.

*

What do we mean by the expression "letting go?" It is evident that on the way to true being, there are many things from which we must inevitably be parted. This is best expressed by Master Eckhart. According to him—and it is an ancient truth—the letting go with which we are concerned is "the letting go of our ego." This movement enables a man to extricate himself from the net in which his identification with his world-ego continuously ensnares him.

This letting go of the ego, however, means much more than merely relinquishing all those objects to which during his lifetime a man has become attached. It entails the giving up of the entire life pattern that has revolved around the "positions" taken by the ego. In this pattern all our thinking, feeling and doing is orientated towards whatever is firmly and finally established. Wherever man is held fast in the grip of any static system (whether of facts, attitudes or values) his life in the world is impermeable to its real meaning. This meaning is always personal and is always in the process of transformation. The only attitude through which it can be discovered is one of inward knowing and existential realization. "Letting go" means forsaking the brilliance of the rational mind and entering what may be called the semidarkness of another form of consciousness in which is revealed an inner light. In this way alone can the rigidity which transfixes all thought and behaviour be dissolved and the kind of thinking that turns all things into objects, give way and die. Only when we let go those attitudes wherein we rely solely upon what we "have, know and can do," will there arise a new consciousness in which the creative dynamism of life is contained.

*

"Letting go" requires not only that we relinquish those imaginings that have determined our understanding in the past, but also the behaviour that corresponds to them. The result of all rational, ego-centered points of view is a kind of wilfulness. Therefore the consequence of letting go must be the abandonment of any attitude that leads us to believe that everything depends upon our own "doing."

An ego orientated towards rational aims is always concerned with notions as to how life really *ought* to be. By letting go in the right way, we learn to "let in" and "let happen" that which, in spite of all our ideas, projections, desires and prejudices, meets us directly in the shape of the world and comes from the constantly-stirring essential being within.

The desire "to do" prevents us from hearing this inner call. By holding fast to conceptual patterns and images, we block the road to our union with Divine Being. Only when we are finally able to give up our ordinary way of thinking will we be able to receive and accept that which is and, as a consequence, give ourselves without reserve to all that arises.

Thus we see that letting go requires that we give up any idea that life must be what *we* expect and *we* desire. Time and again our faith is destroyed when the process of God's justice—inevitably incomprehensible to us—does not correspond with *our* conceptions as to how things should be. Therefore, the prerequisite for genuine faith is to give up all such conceptions. True faith develops only when life is accepted in whatever way that providence, always inconceivably, offers it. It is brought to birth in us when we are able to receive, from unfathomable life and absolute Being that which, through all conditioning, strives toward the light, in the form of our inner states, challenges and impulses.

*

As long as the ego, trapped in its fixations, dominates our lives, all chance of transformation is inhibited. The state of the ego, at any given moment, is manifested by us and recognisable to others by the way in which we are present. However, the process of letting go the faulty attitudes so revealed is by no means a purely inward process. It consists in relinquishing and, as it were, "melting down" the incorrect physical postures which are themselves the incarnation and expression of the world-ego's desire for security. When this letting go occurs it will be obvious to the most casual observer from the way in which a man drops the hunched-up shoulders that speak so strongly of the safety-seeking ego, loosens the lower jaw where the powerful and arbitrary self-will is established, relaxes the tense brow wherein the ego's gaze is transformed into a fixed stare, withdraws the rationally determined eye of the world-ego in favour of the receiving and accepting gaze of the inward eye, and releases the belly (which when pulled in, cuts him off from his basic energy) and makes of it a broad, fluent and firmly rooted body-center able to swing freely in the pelvic area. It should be remembered, however, that in trying to overcome faults in posture we are not concerned merely with cramped "bodily" states, but rather with the manifold manifestations of a person who is inhibited at the center, caught in his world-ego and unable to trust in life. For this reason, if muscles are merely technically relaxed, letting go can never be accomplished in any lasting sense; still less when, towards this end, use is made of injections and massage. It succeeds only when man learns to let himself go in his *ego*, which means to give himself trustingly to whatever comes. Wherever a particular tension makes itself physically felt, this is due to the fact that the relevant individual, through lack of proper trust, holds himself in and becomes tense. Letting oneself go requires, above all else, the inner conviction which enables us to feel, even if we relinquish our world-ego, that we will by no means fall into nothingness. On the contrary, we will be upheld by a state of mind which frees us from relying solely on ourselves and our rational mind, and helps us to live from essential being and to participate in Divine Being. When a man has learnt to yield to his essential being, he has overcome dependence on and fear of the world.

*

The significance of letting go for what may be called our "total posture" lies in giving up that center of gravity which keeps man prisoner in the upper, ego-region of the body. As regards breathing, the importance of letting go lies in the change of emphasis from "doing" to "admitting," from "taking in" to "giving in." In place of the "high," predominantly chest breathing which separates us from the true center, there arises the natural diaphragm breathing, which at one and the same time liberates life and allows it to manifest in the true center.

Every form of letting go has a cleansing and transforming effect, but the reader should be reminded that this is neither a purely technical, physical exercise nor a wholly inner psychic process. It is a personal gesture that, in an analytical sense, is psychologically and physically neutral. At our present level of understanding it is all too easy to underestimate the significance of work on bodily attitudes and to look upon it simply as a technique for increasing the sense of bodily well-being. Such a conception quickly leads to a sort of wrongly-interpreted Yoga and falsifies the effort by allowing it to become mere physical gymnastics. Properly practiced, however, that which from an analytical point of view may appear as the relation between inner and outer processes, is here seen as a twofold expression of the one —the Person.

3. *Becoming One with the Ground*

What significance for us, as the Wheel of Transformation turns, has the phrase "becoming one with the Ground"? We have seen that letting go means relinquishing the forms and structures which in the conditions of our life in space and time have gradually evolved. These conditions are the means whereby the worldly personality, revolving around fixed points and prompted by its desire for security, maintains itself by working simply for itself or the community. To the extent that these conditions become crystallized, they obscure the vital Ground. Man feels himself to be burdened with unused energies and impulses that, because they have been repressed, recoil upon him, and may even paralyse and destroy him.

*

Psychologically understood, the transforming Ground is, in terms of human existence, the "realm of the mother." Any unresolved relationship between child and mother causes an obstruction in this sphere and hinders development. The effect is the same when the mother denies the warmth necessary for the child's growth, or on the other hand holds on to him with a devouring, possessive love. In the first instance, when the child becomes a man, he is inhibited and depressed because he rejects the protests that rise from his depths, defending himself against them by thrusting them down into the unconscious; in the second, he dreads the power of all that embodies the Great Mother in the world for fear it should devour him. When the Ground has been thus blocked, he is robbed of the powers that reign within him for the purpose of redemption, protection and transformation. Thus he becomes wholly dependent on his world-ego, experiencing his repressed depths as frightening forces that either inexplicably attract him or demonically turn against him.

On the other hand, the Ground is also the bridge by which we are brought to experience the maternal aspects of Transcendental Being—that which nourishes, sustains, re-moulds and liberates all that exists in the world. Thus, whenever man is cut off from this maternal Ground, he is always, at the same time, barred from the alternating rhythm of "death and becoming" which itself is a part of life.

*

In all ages Divine Being has dwelt in man as plenitude, order and unity. Of old it was revealed, as man believed, in the power, wisdom and love of the Gods. But today, we are beginning to realize that Divine Being is within the sphere of man's own *experience*. This is particularly true for those who, in fear and despair, have come up against the limits of the strength and wisdom of their world-ego; and for others who, lacking the support of communities that are disintegrating, as once the old Gods disintegrated, are thrown back on themselves.

Just as Divine Being, striving to manifest itself, is present within us, so the experience of union with it becomes at the same time an experience of our *own* being. And, equally, the experience of our own being is an experience of our oneness with Divine Being. If we have once become conscious of ourselves in our essence, we have become conscious of our union with Transcendental Being. We must, therefore, search for that Being which liberates us and inwardly determines our form, and thus discover within ourselves an "immanent transcendence" far beyond the frontiers of our little ego. We shall not succeed in finding it, however, until our world-ego carapace has become permeable. On the other hand, our ability to do without the scaffolding in which we are upheld by our world-ego depends on the strength of our contact with Divine Being. So it is an endless cycle. It is only to the extent that we learn to dissolve the rigidity of established forms and ingrained ideas of how we should be and act in the world, that our essential being can lastingly awake within us. Letting go requires us to withdraw from all theoretical and practical positions by which, on the level of our worldly existence, we keep up the struggle against nature and destiny. When this struggle deprives us of our essential power, we are bound to attempt to free ourselves from our habit of rational classification—although in an intellectual sense this may well be indispensable. It cannot be denied that a proper understanding of the values that are accepted by the world in which we live is a necessary stage on the way to becoming a Person. It is, in fact, the means whereby the egocentric, natural pre-personal ego with all its violent drives, is made to submit to order and custom. It teaches man to conquer his wilfulness through service to something above him. When, however, the ego gives itself over to any paternal authority, to the community, to habit, religious teaching or work in the world, the effect often goes beyond the control of the primary ego (i.e., the as yet uneducated ego one finds in a baby) causing the individuality of our true being to be repressed. Such service and obedience may hinder and limit the immediate uninhibited contact with true life. From this point of view it is clear that the so-called reliable man, the "strong personality," is by no means also a constantly maturing Person or one permeable to Divine Being.

*

In order to become a Person it is necessary to be completely open to life and to accept and integrate essential being. But before integration can occur, essential being must awaken, come forth and outface the difficulties of coming to terms with the world, whether the world be friendly or inimical.

Our discovery of essential being as a reality within ourselves entails also our concurrence in its demand that it become a reality in the world! Essential being is the unconditioned, creative, motivating power of human existence which demands to be given form and which no amount of conditioning has the power to stifle. For this reason, man, finding himself imprisoned in those ingrained habits that prevent essential being from entering his consciousness and coming in contact with the world, is brought to the utmost extremity.

When his world-ego is estranged from Divine Being (even though his essential being is inevitably linked to it) man sooner or later experiences not only the suffering inherent in such estrangement, but also the mysterious tugging of the cord that joins him to his hidden, inner self and from which he is never entirely free. To the extent to which he separates himself from his ego, he will experience this tugging as a painful longing for deliverance. He knows himself summoned by conscience. Something is lovingly drawing him home. Our answer to this summons can only be the determination to work on ourselves in the right way.

The more obdurate and unyielding our concepts and attitudes have become through our conditioning in the world, the harder will it be for essential being to emerge. Instead, its repressed energy will be transformed under such circumstances into a dark force which, in the shape of the Shadow, acts destructively in and through our unconscious. It is, of course, necessary that man should be ethically disciplined. If, however, this results merely in the erection of some ego-superstructure and does not bring about true transformation, it will also cause repression. Man cannot achieve freedom until he cleanses the Ground, recognizes his repressions, and by directing them into fruitful channels renews his acceptance of the powers of the deep.

*

It should be remembered that the ascent to the bright peaks of true being is always preceded by a descent into the dark depths, i.e., into those areas which have been thrust down into the unconscious. It is only when we perceive and accept the Shadow, and painfully recognize our erroneous responses to life; when we perceive, assimilate and live through the (archetypal) figures which personify the repressed energies (the devouring mother etc.), that the way towards true union with essential being is opened and the new self, born from essence, becomes free to grow. All such efforts are fundamentally an expression of obedience to the summons which bids us find the way from our state of alienation back to our beginnings—in other words, to enter into the Ground in order to emerge remade in accordance with essential being. It is a paradox, however, that we can only discover our true home by realizing the extent to which, in the world, we have become estranged from it. But when, by contrasting it with our life in the world, we actually experience our essential home, we are in the position freely and responsibly to bear witness to it wherever, according to destiny, we may live and work among the structures of the I-world-reality.

*

The chief objective of critical awareness is to recognize how and under what circumstances our state fails to correspond to our essential being and to perceive that the faulty state is inevitably reflected in faulty attitudes. The purpose of the practice of letting go is to free the way to essential being; and the purpose of the third step in the process of transformation is union with the Ground. There are many stages and degrees of union. The highest is that in which we are united with the maternal oneness of Being. Here, whatever is inimical to essential being, or has become static and rigid as a result of the workings of the world-ego, is melted down and recast, in order that the new ego may be born. All lesser stages of union should be measured against this ultimate stage. From the standpoint of this, as we may call it, Personal Anthropology, all levels of development should be regarded from "above"—that is to say, from the highest level capable of being achieved by any human being. Man is intended, as his ultimate condition, to reach such a point; his true nature is, in fact, always striving towards this possibility. Even those systems of thought in which man is considered, genetically speaking, as at the "bottom," i.e., in an undeveloped stage, are deluded if they are not orientated towards the highest level for which man is destined.

*

When we speak of Transcendental Being and the transforming and liberating unity of the Ground, the question inevitably arises: is it not presumptuous to inquire into that which so obviously touches the deepest mystery of life and of man, and which is beyond all rational comprehension? Indeed, such inquiry is permissible only if approached with the utmost delicacy. Even so, we should not be so diffident before this great mystery that our diffidence makes us blind and deaf to essential being. The time has now arrived when man has sufficiently matured to experience a reality which has hitherto belonged solely to the realm of faith. Impressions that come, so to speak, from "elsewhere," stirring us to our depths —rousing, enriching and transforming us—enable us not only to believe but to know without doubt that we participate in a level of Being that is "quite other" than that of our familiar world existence—and which we call "transcendental." By means of such experiences we realize that in the secret core of ourselves we are far more than we are normally aware of being. In particular we discover that our unconditioned essence always surpasses what we in our worldly presence—created under abnormal conditions of existence—can ever become. Even so, we have, happily, the right to remind ourselves that the mystery of the Ground of Greater Life constantly penetrates our little personal life, with its continuous summons and its power of renewal. This can sometimes occur with such intensity that we wonder, amazed, how it can be that the restrictive powers of the ego are so often able to hide true Being and repress such mighty energies. Unfortunately, however, for the sake of its own survival and its perennial preoccupation with external facts, the ego continues to work its will within us. Even so, Divine Being stirs in our darkness, bestowing on us, when at last the depths begin to stir and the veil of conceptualization falls, its illuminating experiences.

When a man goes out to meet these intimations of Being and, faithfully pursuing his practice, breaks through the barriers of the system in which life habitually presents itself to his thinking and acting world-ego, then he is at the point of entering "the

Way of the soul's initiation." Our generation is witness to a decisive turning point in man's development—the opening of the initiatory Way, formerly the exclusive province of the mystics, to all those who have arrived at this level of maturity.

*

Once the third step—Union with the Ground—has been taken, then in the language of essential being, Divine Being arises in its threefold aspect: as *plenitude*, whose strength from then on sustains, supports, nourishes and renews a man no matter what his difficulties; as *order*, from which there emerges a hitherto undreamt-of purpose that obliges him to create himself anew; and as an all-embracing *unity* through which he becomes aware and rests upon the world of "one-ness" beyond this world. Such an experience making, as it does, all things new, sets man once and for all on the Way and transforms him from *subjectum mundi* to *subjectum Dei*—which is the "Great Experience," [3] the "Breakthrough to Essence." [4] We cannot achieve it by willing it, nor can we intentionally bring it about; all we can do is to prepare ourselves for it. This preparation is the chief aim of all man's existential self-exploration and practice, and by performing it he is committed to becoming, above all else, a servant of Divine Being.

*

[3] See Dürckheim, *Im Zeichen der Grossen Erfahrung* (The Great Experience), O. W. Barth-Verlag.
[4] See Dürckheim, *Durchbruch zum Wesen* (Breakthrough to Being), Huber Verlag, Stuttgart-Bern.

Many lesser experiences of contact with Being may precede and act as pointers toward this "Great Experience." They are like lights on the Way which lead towards the goal. But, in fact, any genuine meditation in which we succeed in letting go of our world-ego, may bring a foretaste of what is to come. It is possible, too, that a spark of the light which pierces us during the Great Experience may blaze up in any life situation to which we fully give ourselves. It can happen at those times when we find the courage to forego all the ego's prejudices, defences and reservations and, no matter what the cost, allow them to die in order that truth may live. To the extent that the process of transformation takes root in man so that he learns at last to let go—affirming and yielding to all that is in his depths—so will the liberating, illuminating, kindling spark of the Great Light shine more and more frequently until finally it becomes a part of the basic stuff of life. This will occur all the sooner if a man also finds the courage to allow the pure primary impulses to arise freely from essential being—that is to say, those impulses which have not yet been classified and appropriated by the ego. If he can do this there will develop, side by side with this, a more stable attitude, a new and loftier way of being in the world. The center and meaning of life will now no longer be in the world but in Divine Being; no longer in the self-seeking, self-preserving world-ego but in the essence that is constantly bringing about transformation. This whole process results in the formation of a *new consciousness*. For, whereas the consciousness of the world-ego is formed from and made effective by such static concepts as rational determination, the need to classify and the habit of thinking in terms of opposites, the growth of higher consciousness (which is centred in essential being) is evidenced by the rising up of that strength which can face the bottomless abyss and accept the contradictions inherent in life while at the same time admitting the transcendental powers. When man is open to all these, he experiences such an increase of strength that he is able to realize his own essential form, even under worldly conditions.

*

It is inevitable that our first approach to Transcendental Being should shake the structure of our ordinary life. Such encounters occur much earlier in life than is generally supposed. "First experiences" [5] may arise in early childhood during the transition from one level of consciousness to another. They often occur at the moment when the original oneness is shattered and the child discovers that his "I" and the world are not one—that, in fact, they are in opposition to each other. The child cannot attempt to grasp the purpose that lies behind this incomprehensible experience, but the grown man feels himself compelled to try to understand intellectually the breaking-up of the old shell that brings him the experience of Being; and this compulsion can have disastrous effects. As long as his world-ego remains dominant he immediately transforms into mental concepts any recurrent awareness of Essential Being. In the same way, this grown man, when sought out by the Divine,

[5] See Gebsattel, "Ersterlebnisse" (First Experiences), in the *Festschrift* for Buitendyk 1961.

will always be tempted, at the moment of encounter —the moment when it might be possible to discover that which in his essential being he has searched for all his life—to label, fix and classify the experience. And by so doing he loses it. Man cheats himself of the benefit of any shock that may have been vouchsafed him, when he tries to reduce its incomprehensible and, for that reason, frightening content, to familiar concepts and images. As a consequence, the numinous core of the experience and the sense of blessedness that arises from it, are lost. If we are to be receptive to the Ground of Being, we need to have the courage to meet the unknown, to renounce the right to cognize and tabulate, and to endure the mystery that which cannot be conceptually comprehended—in short, to pause and inwardly dwell in that to which we are all to unaccustomed, the radiance of Divine Being.

*

The practice of becoming one with the Ground— the third spoke in the Wheel of Transformation— requires above all, practice in learning to endure the awakening of Essential Being in the conscious mind. Therewith a new dimension arises which is forever inaccessible to the world-ego. Everything that serves to undermine the foundations or to destroy the supports on which the world-ego has hitherto relied and established itself implements the exercise that prepares us for this kind of inner knowing.[8] Not only

[8] See Dürckheim, *Zen und Wir* (Zen and Ourselves), 1961, O. W. Barth-Verlag.

does it now become necessary for us to question every apparently secure and consolatory opinion; the ego's fear of pain and annihilation—which arises from its concern for its own survival—has also to be conquered. It is only when man has learned to risk over and over again all that has seemed assured, that that which is forever unknowable breaks luminously upon him. Only then can Greater Life bestow on us its light as well as its darkness, only then can Divine Being enfold, renew and transform us.

*

The man who, being really on the Way, falls upon hard times in the world will not, as a consequence, turn to that friend who offers him refuge and comfort and encourages his old self to survive. Rather, he will seek out someone who will faithfully and inexorably help him to risk himself, so that he may endure the suffering and pass courageously through it, thus making of it a "raft that leads to the far shore." Only to the extent that man exposes himself over and over again to annihilation, can that which is indestructible arise within him. In this lies the dignity of daring. Thus, the aim of practice is not to develop an attitude which allows a man to acquire a state of harmony and peace wherein nothing can ever trouble him. On the contrary, practice should teach him let himself to assaulted, perturbed, moved, insulted, broken and battered—that is to say it should enable him to dare to let go his futile hankering after harmony, surcease from pain, and a comfortable life in order that he may discover, in doing battle with the forces that oppose him, that which awaits him beyond the world of opposites. The first necessity is that we should have the *courage to face life*, and to encounter all that is most perilous in the world. When this is possible, meditation itself becomes the means by which we accept and welcome the demons which arise from the unconscious—a process very different from the practice of concentration on some object as a protection against such forces. Only if we venture repeatedly through zones of annihilation can our contact with Divine Being, which is beyond annihilation, become firm and stable. The more a man learns wholeheartedly to confront the world that threatens him with isolation, the more are the depths of the Ground of Being revealed and the possibilities of new life and Becoming opened.

*

The encounter with essential being which sets us free from all the bonds of the ego, and the awareness of Divine Being which puts an end to all worldly attitudes, are joyous and liberating experiences. But the process of becoming one with Being, through the dissolution of all that opposes true life, should not be thought of as an end in itself. When the world-ego is, as it were, cancelled, the experience of Divine Being must not be allowed to ebb away. Rather, it must be converted into a creative impulse towards a new form. Only thus is Union with the Ground of Being fulfilled. The whole point of the painful tension between world-ego and Essential Being is that by this process the secret meaning hidden in this tension becomes clear. In a word, it is to lead man from the wrong way to the right way—not to bring him safe home to some kind of "eternal resting-place," but rather to draw him on towards the painful, rewarding, never-ending process of transformation. By this means alone can he approach the fulfilment of his destiny—which is to become a Person in whom and through whom Divine Being is revealed in this life.

4. New-Becoming
(The inner image and decision)

I remember the case of a woman who, being seriously ill and, as she believed, dying, felt herself finally entering into the Spiritual Ground. It was a blissful state of being, she said, in which she was absorbed into a boundless sea of love. She remembered, too, how joyfully and unresistingly she felt herself "passing over" into this state of all-redeeming infinitude. And then a strange thing happened. Suddenly, in the midst of all the brightness, she became aware of herself as something distinct from it, a shining nucleus. At the same moment she knew that she was not to be permitted to depart, but was required to return to life. The feeling of being herself a nucleus of light coincided with the impulse—even, she felt, the sense of obligation—to return to the world. And this impulse had, at the same time, such authority that she could not but obey. So she remained alive.

This incident shows very clearly the essential elements that characterize the dawning of New-Becoming. These are: the entry into the ground, the encounter with the nucleus and the making of a new start. The purpose of these stages is clearly revealed in the great experience. The principle of these three steps governs each correctly accomplished exerci-

tium (even that of meditative sitting). It also applies
to every situation in daily life that is carried out according to the demands of our essential being.

*

The Great Experience[7]—which leads to Metanoia,
to a reversal of the old life and to new birth—contains two elements: the experience of oneness, in the
stillness of which all things are rendered down, and
the encounter with the special way in which we as
individual beings participate in Being as Greater
Life, which drives forward with dynamic force into
the "light of the world." Our experience of Divine
Being as the Life which strives within us to manifest
in the world, is none other than Essential Being. The
Great Experience necessitates first the melting down
of the world-ego structure and this, when it happens,
has the effect of liberating a man from his hitherto
compulsive obligations towards it. The second requirement is that man must encounter his own essential being and experience its drive to become
manifest in a certain form, and thus arouse his creativity and his sense of responsibility towards life.
Liberation from the ego allows us to take the step
that leads to the birth of the Person. The birth itself,
however, occurs only when the experience of Essential Being is felt as a compelling summons. Thus we
can see that the encounter with our essential being,
which presupposes release from our old selves, does
not merely bring us to the blissful awareness of our

[7] See Dürckheim, *Im Zeichen der Grossen Erfahrung* (The
Great Experience), 1961, O. W. Barth-Verlag.

own individuality, or a feeling of jubilation at personally participating in Divine Being, i.e., "that one *is* oneself *it*." It has the further effect of reminding us of the heavy responsibility of our human mission, which is to give evidence of our participation in Transcendental Being in terms of our own individuality in the life of space and time. Once we open the door to Essential Being, we are required to make an irrevocable decision to be answerable for the realization of its image in a form that is viable for life. Furthermore, we must henceforth, in each fresh contact with the world, affirm our relationship with Essence and bear witness to the fact that we are indeed that "someone" we knew ourselves so absolutely to be in that first real encounter with the Great Experience. Indeed, that which may be experienced in some propitious moment as the Great Experience, may resound through every contact with Divine Being, no matter how slight it may be.

*

The summons to realize and give proof of our essential being amid all worldly conditions is an absolute demand. It refers not only to our own development, but equally to the ordering and shaping of the *world itself*, with which we are so closely linked. But while we are absorbed in mastering or serving daily life, it is essential that at the same time we bear witness to Divine Being. Thus, in the final instance, the purpose of our service to the world lies not solely in our personal aims and personal salvation nor solely in the perfecting of the world, but in our service to both. We are, in fact, intended to take up the cause of vital Being both within ourselves and without. In the course of the Great Experience man discovers that in his essence he is destined to become a particular "someone." At the same time, he becomes aware of the nature of the task he must undertake, the task of fashioning himself and the world to a specific requirement. The particular character of that form which it is man's and the world's destiny to achieve lies in its transparence towards Divine Being.[8] It is only through the "becoming-transparent" of the created world that that which a man has experienced as an inner image and sensed as a call to the Inner Way gains concrete reality for him. The human mode of this corporeal reality of the inner image of the world is the Person. This can never be a static form, for it must participate in the endless process of transformation and by this very process become transparent. Here we have a "formula" for life, a means by which, if man accepts the ceaseless alternation of "death-and-becoming," he will more and more experience Divine Being as plentitude, form and oneness.

*

For the perception of the true self, for its evolution and the power to put it to the proof (as well as for the proper guidance of others) there are three essential requirements: a capacity for experiencing the inner image as the manner in which we and the world are destined to participate in the never-ending

[8] See Dürckheim, "Auf dem Wege zur Transparenz" On the Way towards Transparence) in *Transparente Welt*, 1965, Huber-Verlag Stuttgart-Bern.

process of transformation and becoming; an ability clearly to perceive and steadfastly to retain the concept of the Way; and the skilfulness, in special periods of practice (as well as in the general practice of daily life), with which we can transform ourselves and our world so as to become continually more transparent to Divine Being.

*

In these days, when the efforts made towards redressing the neurotic disorders that prevent man's ability to work and to make contact with himself and the world are teaching us more and more about the way he functions, it is to some extent natural that the main emphasis should be placed on the significance of his personal history. Such emphasis, however, should not be allowed to blind us to the true image of man or to the law that governs his Way. Apart from his personal history, he is required to manifest super-historical life in the historical world, and in this lies the real purpose of human existence. There are few today who properly value, either in theory or in practice, the innate image of man or his inward Way—and neither of these aspects can be reduced to psychological or sociological elements.

*

What do we mean by the term "innate image"? Let us suppose, say, that we come upon a drunken woman. Why does such a sight shock us? It is not because she is violating "the essential image" of a human being? Is it not this that offends us, quite apart from the life-story or the particular life-situation that might explain her drunkenness? Such a spectacle can sometimes so deeply repel our sense of "inner propriety" that though we may understand and even sympathise with the woman's situation, we are outraged by such a violation of human dignity. The extent to which this dignity is apparent in any human being is evidence of an attitude of fidelity to Divine Being. Such an attitude is dependent on the degree of his or her ability to keep faith with that which we have called the innate image of man.

The innate image should not be thought of as a mere idealized or abstract conception. On the contrary, it is the basic reality of our human existence. By its power to shape life, to awaken conscience, to bid us harken to the summons that calls us to follow the individual Way, the image we bear within us proves its reality. We could even say that fundamentally the innate image *is* our "innate way." We are continually driven to find a manner of life that corresponds to and manifests it. If this drive in man is repressed and its demands unheeded, the result will be a harmful inner condition and even physical illness. The innate image of man is indeed no other than his essential being, understood as the driving force, the feeling of obligation and the fundamental longing which determine the particular pattern of growth through which Divine Being may be revealed. According to this pattern, which is not solely biological and is quite independent of the external conditions of life, a definite succession of stages leading to the state of transparence is sharply delineated.

*

The innate image is the inborn Way that leads to our real Self. This Inner Way has absolute validity for all men—not merely for the particular life of the individual. Its realization can never be dependent on circumstances whatever they may be. Its inescapable demand for realization proves the reality of the innate image and when man ignores or represses it he becomes physically or mentally ill.

The need to be given its due, which is inherent in the innate image, is experienced by man as an immanent summons to realize himself and his world according to the pattern appropriate to his particular nature. For this to be possible, man's state of mind and body must be such that it enables him to act in harmony with essential being under all circumstances. No special pleading arising from difficult life situations has any validity here. It follows therefore that in the presence of a man who is suffering the consequences of having failed himself, a mere theoretical understanding of his situation is quite useless: he will need something more.

*

In these days, when our insight into the psychological conditions of human development is becoming increasingly enlarged, we are able to recognize more and more clearly the origins of the distortions in behavior. Such discernment, however, can easily mislead us into becoming all too "understanding" of conditions that have arisen of necessity from the circumstances of space and time. If we take the line that as well as perceiving the origin of a man's error, we must also excuse it, we may well be transgressing the inherent law of his personal growth. By taking such an attitude we will inhibit the hidden yet ever present need in man that the absolute claim of his essential being be taken seriously and he himself shown the way to its fulfillment. In offering only "understanding," commiseration and forbearance, we may be by-passing a man's chance of freedom and his possibility of achieving personal responsibility. In the mistaken belief that we are helping him, we may in fact be barring the Way of Transformation to him.

It can happen that a man who has missed the Way towards the realization of his essential self may so suffer as a result of this that he is brought closer to spiritual birth than someone who has been denied this special kind of anguish. That man who feels himself lost in utter darkness in the world which, so long as he is caught in his ego, thrusts him into fear, despair and loneliness, may be the one uniquely ready to hear the call of his essential being—ready to respond to the summons that, breaking through his ego-shell, brings him to the awareness of his inner core. Thus, when someone entrusted to our care has fallen, perhaps, into utter despair and reached the limits of his alienation from Divine Being, it may be necessary for us to find the courage to deny him sympathy and consolation. Bearing in mind his innate image and our own allegiance to truth, we may rather need to call him to enter upon his innate Way in order that his essential self may begin its struggle towards the light. This is the whole purpose for which his state of darkness was given him. But our attempts to remind him of his true self must not be an authoritative "regimental" call to order. By protesting against such a command his ego would only

be further strengthened! We should rather appeal to
him through the medium of his innate image and so
call him back to his essence. A real man would prefer
to suffer the consequences of being faithful to Essential Being, rather than buy freedom from this suffering at a cost of betraying it. In the long run it is disastrous when a man's health and efficiency, lacking
contact with Essence, do not correspond to his true
nature, or when his easy adaptability to the world—
by seeming to make such contact unnecessary—hinders it. On the other hand the very fact of being
confronted with the innate image releases a miracu-
lous, all-transforming power. We may therefore confidently trust in the healing force of the innate image
and rely on it not only to dissolve the painful rigidity
of our mistaken response to life conditions, but also
to release the creative power of the Ground.

*

The emerging and unfolding of the innate image,
and the process of becoming one with it, is not only
an inner experience: the body itself undergoes transformation at the same time. We can, in this respect,
speak of "the corporeality of the innate image." But
what do we mean by this expression? It conveys the
fact that Essential Being, through practice, becomes
apparent within a man's physical life-form (*Lebensgestalt*). That is to say, Essential Being is manifested
in the manner in which he is present in the world.
The permeable, ever-changing form which man acquires as he progressively brings to birth his inner
image, is very different from his inflexible, impermeable I-world-form which, being born out of de-

sires and fears, indicates his mistaken drive for security. Different again, because it has not grown from Essential Being, is the Persona, which represents a conception of the perfect form but in doing so buries the Shadow.

It is as little possible for a perfect form of the innate image to exist as it is for a perfect, so to speak, form of life conditioned by the world. In the latter case, even where adaptation to external situations has clouded Essential Being, the innate image still fitfully shines through; and in the former, even under the most favorable circumstances the appearance of the innate image is always to some extent conditioned by the world. A valid life-form (*Lebensgestalt*) depends on the right relationship between both. In every approach we make toward understanding and guiding men, these two aspects should be kept in mind. In the process of becoming, the predominance of an orientation towards the world is inevitably succeeded by one which gives precedence to Essential Being. Only in this way can the demand which arises from within be fulfilled and the never-ending inward journey faithfully followed.

*

Our life-form, which begins to emerge during the process of growing-up, is largely a result of our confrontation with all the historical circumstances of life. In seeking to come to terms with these factors man tries to conduct himself in such a way as to enable him to adapt as painlessly as possible to the demands, opportunities and dangers of the world.

When it happens that a human being grows up

under difficult circumstances, or from lack of love and understanding inevitably becomes distorted, he is thrown back upon himself. Inevitably, he reacts to such conditions by developing postures which are the result of continuous adaptation. By means of these he protects his natural ego, but always at the expense of the growth of his individual essence. Such forms of adaptation hinder the arising of the vital strength that lies at the very root of himself. Furthermore, they injure his essential drive to acquire individual form (*Gestalt*) and reduce his capacity for love. The more rigid such habits become, the more does his confidence increasingly depend on his own efficiency, his possessions, his intellectual knowledge and all his worldly capabilities. Thus his reliance on the outer world continues to expand, and as a result he becomes the prey of loneliness and isolation. If it should happen, however, that he finds the way back to the fundamental energy of his essential being, then the awareness of his own real strength and innate worth helps him to become again independent of the world, while at the same time remaining inwardly united with it. For through his essential form, Divine Being, though it transcends the "world," penetrates and links together all aspects of man's existence.

The man who, as a result of being in accord with his innate image, is genuinely present, impresses all who meet him. Representing, as he inescapably does, their own embodied conscience, he becomes for them a mirror in which their own search is reflected. For within every man lies the hope—and the wish— that he may not only be described but challenged by his best elements; this is true even when, as a result of unfortunate circumstances, any distortion in himself might seem to be justified. Fundamentally our longing for essential self-realization is always stronger and deeper than the desire to be acquitted of our shortcomings.

*

From the moment in which man encounters his innate image and hears its summons, it becomes necessary for him to heed the call. It is not enough to have been momentarily seized by it. Of our own free choice we must also affirm and decide freely to accept the impulse from Essential Being that has taken possession of us and this requires that we in turn consciously seize upon it. We must renounce the self-will of our worldly-ego and accept instead the responsibility for that which comes to us from the fundamental Ground. It is a misconception to suppose that, in obeying Essential Being, man is giving away his liberty and his will. On the contrary, only by submitting his world-ego to the demands of Essential Being does he truly acquire that free will he is intended to possess in his relation to the world. This worldly will itself becomes productive and mature only when it is put at the service of a wish that does not stem from the world-ego, but from the experience of the transcendental Ground of Being.

*

To return for a moment to the Wheel of Transformation, it should be said that each of the five steps which together compose it may at certain moments in our lives be experienced with particular intensity.

For instance, in such moments a man may suddenly become aware, with a feeling of anguish, of some longstanding wrong attitude he may have held towards his nearest and dearest. In this realization he may vividly experience a state of self-critical awareness. Or again, he may realize that some blow of fate has cracked the shell in which he has been enclosed and flung him into a situation where he finds himself letting go of everything he had hitherto clung to with all the strength of his being. Again, it may happen that he finds himself so inwardly shaken that he is carried beyond the narrow limits of his human understanding and capacity, and is thus enabled to break through to the core of his true self. Quite unexpectedly he may find himself brought into contact with Divine Being, the way to which had hitherto been obstructed. Finally, in some propitious moment, he may encounter his essential being in such a way that he is instantly freed from that life-form, wherein in his adaptation to the world, he has betrayed his own being. Thus, transformed by such an experience, he will feel the responsibility to find the life-form (*Lebensgestalt*) that conforms to his innate image.

At any time in our daily life it can happen that we find ourselves secretly searching for, and even unexpectedly coming upon, that which, in some unique past moment, we once deeply experienced. It should be remembered that—through practice—the possibility exists for us always to be in a state of readiness for this. Once we have really grasped the purpose of our life and committed ourselves to the Way of Transformation, everything that happens becomes an occasion for bringing about the right state of mind. To this end we need to devote ourselves to the continuous practice of special exercises (*exercitium ad inte-grum*).

*

One fundamental exercise, which incorporates all the elements necessary for the process of transformation, is the practice of correct, that is to say, upright sitting. Throughout the Eastern world—from India to Japan—meditative sitting is the foundation-stone of all spiritual practice. Such sitting, however, should not be thought of as being the prerogative only of Eastern man. There is no one, in the East or the West who, critically aware and practising correct sitting, does not very soon realize—by observing his posture, tensions and breathing—that there is something wrong with his way of existing. This realization goes far beyond mentally noting his physical distortions. Its prime purpose is to make a man actively aware of his wrong attitudes, and his whole way of being in the world. By abandoning those tensions in the upper part of the body that indicate the ego's desire for security; by letting oneself down into and becoming one with the center (Hara); by sensing and admitting the true, upright form which arises from contact with this center, and finally by attempting to give proof of it in the wider context of everyday life—all these processes, through the medium of correct sitting, become opportunities for the practice of a right attitude towards every aspect of life.

All too often I find someone sitting opposite me in a posture that inevitably robs him of all his central strength. The sunken chest, the bent head, the drawn up knees, the concavity of the torso, are all characteristic of a man who, as *himself*, is not really present in any sense. If I should then say: "What kind of posture do you call that?" inevitably there would be protest. Young people, in particular, like to make it clear that they are not in favor of the "sit-up-straight" attitude. It reminds them unpleasantly, they may say, of an authoritarian father. This is certainly understandable. But if one then asks: "How do you *prefer* to hold yourself?" the reply is usually, "I just like to sit comfortably!" It takes no more than a few minutes to show such a person that—between one posture in which he is unnaturally drawn up, stiff as a ram rod, and another in which, losing all sense of himself, he collapses like a folded umbrella—there is a third possibility. In this third kind of sitting he is able to be present in his center in such a way as to be both upright and balanced, freed from any rigidity that could do violence to his form and at the same time insured against the kind of collapse which pulls him downwards towards dissolution. In this new posture he will soon discover that he is both at ease and rightly ordered and—what is more important— that he is present as *himself*.

This is what a man experiences when he gives up his false ego-posture—an attitude that is as much apparent in rigidity as it is in dissolution. He is now able to let himself rest serenely in his basic center, admitting (and permitting to enter his will) that which comes to him as the right form. It is worth noticing that whenever a man's shoulders are hunched up and his center of gravity displaced too far upwards he is as significantly without a true center as he is when he sinks down into formlessness and inertia. Whether it be correct or incorrect, every attitude is the result—conscious or unconscious—of a man's own will. As we all know, an animal can sleep standing up, but if a man falls asleep while standing he very soon collapses. It is incumbent on man, therefore, to be constantly watchful. Not for a moment is he absolved from responsibility for the way in which he is present at any given time.

*

Correct upright posture should not be thought of as having anything to do with being "puffed-up" or taking on the extravagant stance that is the mark of an artificial persona. On the contrary, it is always an expression of true humility. There is a certain modesty about correct posture in that it rejects anything that surpasses its own measure, whilst accepting and revealing unreservedly as much as corresponds to that measure. Therefore, to remain below our true level is false modesty; and false modesty is as ego-centered and existentially untruthful as the pretence that we are something we are not. It is clear, therefore, that correct posture has as its primary requirement that the innate image be allowed its appropriate form. Only under such conditions can we be sure that is it we ourselves who are really present, ready and able to respond to life without either fear or presumption. Here is a case in point.

Some time ago, a young Hungarian who had been hard-hit by life came to consult me. Up to the age of fourteen, while growing up at boarding-school, he built up such an ideal image of his long-absent mother that she became for him, in his mind's eye, the most perfect of loving parents. However, when he met her in the flesh, he was shocked to find that she turned out to be—for him at any rate—all that he most hated. The force of his disappointment plunged him into the role of the perpetual avenger, of one inwardly compelled to destroy what he loves and to devalue and debase those things that others hold sacred. Together with all this, he was immensely talented, a successful designer and actor, and a fully trained dancer. It was this last factor—his dancing—that made it possible for me to yield to his request that I should, without insisting on any preliminary training, teach him the practice of "Hara." So, I undertook to show him how to come to his true center. The process of letting go was no problem for him; his difficulty lay in standing upright correctly and allowing the true form that corresponded to his innate image to emerge. At last, however, he was able to do this correctly, and I left him standing there, in the middle of the room, letting him talk to me while I myself sat down. I used this as a test in order to discover whether, when left to himself, he would immediately fall back into his old destructive rebellious attitudes. In fact, the opposite occurred. Since he was in the "right posture" it did not wound his feelings to be left standing. Indeed, to my astonishment, he suddenly, but politely, interrupted my conversation by saying: "I must tell you something. A very strange but, I believe, important thing is happening to me. At this moment I am experiencing, for the first time in my life, what is probably called humility." And now, I saw for the first time the light of truth in his eyes.

This, of course, was only a "first experience" and such an occurrence by itself does not signify transformation. But the first is, nevertheless, always a decisive experience, and in this case it had a tremendous effect. For this young man, at the moment of experiencing his own truth a new life began. This is an example of how, through the assumption of a right inner attitude, a liberating impulse can emerge. The healing power of this impulse comes from the fact that the innate image is simultaneously realized and manifested by the "pure gesture." This, of course, can also happen during practice, without, as in the case above, being related to such difficult life problems.

5. Proof and Practice in Everyday Life

If a man is to grow into a Person it can only be by repeated resolutions of the painful contradictions between the transformation which the law of his essential being requires of him, and his desire to cling to those acquired methods of adaptation that seem—though mistakenly—to set him free from friction. Even when a man has reached the highest possible point of his development, he is able to respond to the claims of super-historical, or absolute, Being only through his historically conditioned and, as it were, outer habiliment. We are never able entirely to divest ourselves of what we have become. None of us can measure up to the image, or form, of the essential being that exists within us. What we can do is to so practise that our personality—which is conditioned by the world—becomes as transparent as possible to that image.

*

Practice may be thought of as having a twofold aspect: firstly, we must at specific times abstract ourselves from the daily round and by correct breathing or meditative sitting practise letting go, in order to achieve oneness and to be made new; secondly, we must live the ordinary day as practice. Each of our daily actions not only has an external import with regard to its inevitable results in the world but, by the manner in which it is carried out, it also reveals an inner significance and provides an opportunity for inner growth. Once a man is on the Way, every act and every task can be the means by which the Wheel of Transformation achieves another revolution. By means of critical awareness of our incorrect postures, by letting go, by surrendering so completely that we become unified, by admitting into ourselves the innate image and demonstrating it in the rightness of our actions, the five steps are accomplished.

In so far as we allow our daily life to put us continually to the inner test we are, in effect, practising the fifth step. Thus, amidst the conditions created by any fleeting situation, we should endeavour to bear witness during the daily round to that which we may have experienced with particular clarity while carrying out special exercises in the quiet and privacy of our room. Practice itself is simple, but it is not at all easy to become *one who practises*. The fact that we may have succeeded in some particular exercise, privately practised, is no assurance that a similar success will be available to us in our life in the world. To realize that an attitude which may have brought us our first contact with Divine Being during isolated practice can so easily be lost, is always a painful experience. This may happen not merely from the pressure of adverse circumstances but is, in fact, the common result of our being "out in the world." Such suffering is inevitable: it comes from our daily failures to permit unconditioned Essential Being and Divine Being to take shape within our historical existence. Indeed, for a human being to achieve this with any degree of continuity, however hard he tries, is possible only to a certain extent.

*

When we look at a tree we see in its form, on the one hand, the manifestation of its innate image, and on the other, the result of the conditions that have governed its manifestation. Certain trees have living shapes that seem to correspond fully to their essential form. There are others whose bizarre contortions— as, for instance, the old weathered beeches at Schauinsland—speak so vividly of the many storms they have endured, that it is evident that their original innate image has almost entirely disappeared, or appears only as a negation of it. And yet, are not these very beeches the most convincing witnesses to what happens in daily life? And does not the very manner in which they deviate from their own innate image actually reveal it?

In the same way, each individual man's worldly form discloses this double origin. But unlike plants and animals, man is responsible for the extent to which his innate image is enabled to reveal itself and develop, no matter under what adverse conditions. However, the fact that this is never more than partly possible must be accepted as an unalterable element of human existence.

*

Sometimes it may happen that in talking with a friend about somebody else the question will arise as to what kind of person that "somebody" really is. This inevitably brings us up against the contradiction —as it appears to us—between his essential being and his empirical personality. In the light of the latter, we may perhaps judge his personality too harshly. But if we measure him against his innate image, making allowances for what he may have suffered, we may perhaps put aside our criticism of what, through worldly conditions, he has become, and feel ourselves bound to admit: "But, after all, he is a real human being!" In making such an admission we may find ourselves taking into account the many vicissitudes of his life and will honor and give him credit for simply and honestly accepting his life-body. (*Lebensleib*)! It may even be that, for this very reason, we feel ourselves drawn to him in human fellowship. We begin to see that by not dissembling his imperfections he gives evidence of the corporeal unity of that which is ordained and made possible for man. In the end we may even come to value him for the fact that he neither pretends to be less than he is, nor wishes to be more. Here we are confronted with a mystery. When a man, trying to do—and to be—what is right, simply accepts what he has become, he thereby grows transparent to his essential form even when he is not fully in accord with it. Thus, as long as he stands fast to whatever is his own truth, his inborn, super-historical being glows like a light in the midst of his ordinary existence. And this can happen even when, under the adverse circumstances of space and time, the life-body has been gravely wounded. Whenever a man struggles to be true to himself and at the same time stands simply and honestly by that which he has become, accepting himself in all his weakness and imperfection, his essential being inevitably shines through.

In the same way that our inborn, fundamental perception of the Way, in which momentary states of being do or do not correspond to our innate image, can be awakened, so we can train ourselves to recognize the genuineness of our attitudes when confronted with what we actually are. By means of constant practice we can develop our awareness of those attitudes that correspond to our innate image as well as strive to become sensitive to the genuineness of our gestures, actions, and behavior. We endanger this ability when, for the sake of some idea—usually false—of how we should be, we dishonestly cover up what actually is. Only extremely rarely do we succeed in the twofold task of developing our awareness of our true attitudes while at the same time refraining from hiding those we feel to be less creditable. For this reason we should respect—indeed, revere— the purity of those gestures and postures that enable the light of our Divine Ground to shine out undisguised, even through the often (to us) painful affirmation of our worldly manifestation.

*

Any man who has really entered on the Way is bound to discover that henceforth he will never again be free, even for an instant, from the responsibility for attesting to essential truth in his attitudes and his life. He must ceaselessly endeavour to fulfil this responsibility, knowing very well that, no matter what he may achieve, it will always be inadequate compared with the demand of the innate image. Above all, he inevitably learns that whatever light may shine forth from him and his works, it is never he himself who has created it—all he can do is receive and accept it, in the sure knowledge that it does not originate in him. The most important prerequisite for any success in our practice is that there should appear in all our actions the awareness that anything we may achieve in regard to transcendence is not done of ourselves. All that can be said is that we have been able for a moment to accept and admit that towards which the Divine Ground constantly impels us. There is blessing indeed to be received but it only comes when, by gathering all our energy into the effort towards self-realization, the compulsive conviction "I must do it all myself" increasingly diminishes. We need to learn to trust and to accept that which reigns concealed within us. On the other hand, when we become aware of our own weakness, the conscience that enables us scrupulously to develop the form that corresponds to the innate image must without fail take root. Recognition of conscience and obedience to such recognition should always go hand in hand, if we are fruitfully to serve the manifestation of Divine Being in our lives. It should be remembered, too, that each time we become aware of falling short of what is demanded of us, and with each new insight into our failures, the fifth step once more becomes the first. For it is through "critical awareness" that we are enabled to realize how and when we exist and behave in a manner unbecoming to one who bears witness to Divine Life.

*

Bearing witness to Divine Being in the right way has as its corollary the continuous turning of the Wheel of Transformation. For this continuous turning two factors are responsible, the most fundamental being the fact that Divine Being, quite without our assistance, keeps us constantly moving. By means of this activity it incessantly works within us. It responds to our every deviation from the form appropriate to it with a more or less gentle homeward pull. We are never entirely set free from our oneness with Divine Being. The longing and restlessness in our hearts and the pressure of our conscience progressively makes us aware of this bond.

The second factor that keeps the Wheel of Transformation turning has its origin in man's free will. It must be remembered that only in the measure that we have unconditionally incorporated the Divine Will into our *own* will, are we able to stay on the Way. Sooner or later the time comes when we have to decide, once and for all, to remain open—not only to everything of good or bad that life inescapably brings us—but also to that which joyously flows through us from the depths of Essential Being and makes upon us its absolute demand. There will inevitably come a time when we must discover in ourselves a state of readiness to bear witness to this under all conditions of existence. For it is only by means of our *becoming* that Divine Being can be revealed in the world. As Master Eckhart says, "Divine Being is our Becoming." However, such essential Becoming can only be manifested by an attitude that embraces every spiritual and physical expression of a man's life. And only this way of being "in form" enables Essential Being to express itself more and more freely in all our postures and gestures, and to strengthen the Person in his progress towards becoming more and more transparent. There is no pattern inwardly alive in man that is not reflected in the pattern of his physical behavior. The more deeply a man becomes aware of his whole disastrous attitude to life and the way in which he lives it—the more surely, as he approaches transformation, will he begin to recognize the extent to which the distortion and rigidity of his postures have become physically ingrained. Henceforth, perhaps, he will undertake to practice those gestures which further that transparence to Divine Being which is preordained for man.

*

Some years ago, a lady came to see me asking: "How is one supposed to pray?" "Can you kneel?" I asked her. My question must have touched her on a sensitive spot, for she retorted irritably, "What do you mean?" "It is quite simple," I said. "You just kneel down by your bed and give yourself up completely. . . ." At that she trembled, rose from her seat and left the room, without a word of good-bye.

The next day she reported the following incidents. After she had left me, she had, she said, staggered as if struck by a blow, then suddenly she had begun to run, faster and faster, until, arriving at her hotel, she stormed up the stairs, locked her door and fell upon her knees. And then "it had suddenly come over her." She did not understand what had happened; all she knew was that in this posture she had been powerless and as it were, extinguished, and at the same time delivered up to something that protected her, and—yes, now she knew what it was all about!

This is an example of the healing power of pure gesture.

*

One prerequisite of our being able truthfully to bear witness to Essential Being in our existence in space and time, is that we must inevitably pass through a zone of annihilation. The notion that it is possible, once and for all, to achieve the ecstatic state of living in the presence of Divine Being, is erroneous. Similarly, the conception of the "perfect man," fully rounded and at one with himself, is a delusion. The tales one hears of men who can no longer be influenced by anything, and over whom the world's opposing forces of light and dark have no power, are misunderstandings, if not betrayals of the truth.

*

Only the man who does not fall into the error of imagining that he is "complete," is able to give proof of Essential Being in the midst of life. Only he who

knows that he will never come to the end of his
efforts can cope with the world as it is—indeed, for
such a one its very insults have the effect of unifying
him still further, to the extent that he is even ready
to take up the gauntlet and do battle with it. What
matters here is that he fights in the right way! [9] And
this depends on the degree to which he is "right" in
himself.

To be right in this sense, it is not enough for a man
who wishes only to defend himself and gain his personal ends, to take up the fight simply from the
standpoint of his pragmatic thinking and all the familiar techniques of his ego. Nor, on the other hand,
is that man right who, in the service of the tradition-
ally established values of his community, suppresses
his ego and makes great sacrifices. Indeed, we must
go further and assert that not even he is right who,
by relating himself solely to the forces of light, ignores and denies the dark forces—those primal powers that are dark simply because they are repressed.
By disregarding these, and failing to give them their
due, he actually calls them into being, both within
himself and without. Only the man who is able to
stand fast and imperturbably allow the darkness
within and without him to approach, making no attempt to evade its menace, can give proof of
Essential Being in life. He is capable of this, however,
only when, in obedience to that law of life which
permits no respite, he risks repeatedly whatever
worldly position he may have won. He must be prepared to test the form that seemed to be

[9] See Dürckheim, "The Right to Fight" in *The Japanese
Cult of Tranquillity*, Ryder and Co., and *Wunderbare Katze*,
O. W. Barth-Verlag, 1965.

comformable with his essential being over and over
again in renewed encounters with the threatening
world. Our task is not merely to conquer the primary ego within us, nor to overcome whatever form,
corresponding to the world's requirements in the
matter of efficiency and social conduct, inhibits our
essential being. It is even necessary for us to hazard
that very state of mind which, from the time
when it began to manifest itself, has grown out of
contact with Essential Being. Without knowing it, all
men, as soon as they have recognized the core of
their essential being, begin to build a temple around
it. But this is the very edifice that needs to be destroyed—not once but again and again—in order that
Divine Being itself may remain alive and renew its
light within us. And for this no temple is necessary—
all that is needed is transparency.

*

Divine Being is beyond all opposites. It is undoubt-
edly present within us, but it cannot flourish in this
life of ours if, ignoring the multiform and conflicting
aspects of the world, we remove ourselves from the
market place and dwell in a place apart. Man can
only grow from the root of Essential Being when he
allows even those things that are repugnant to approach him. He must without reservation confront
the powers of the world just as they are, neither
avoiding the dark, nor lingering in the light. It is
only by freely and repeatedly choosing new encounters, by marching on and, when necessary, yielding
up that which has been most dearly bought, that the
skin, so to speak, of the inner man (which is neces-

sary for his survival in the world) can develop and
grow strong, and the instructions needed for the
building of a new, more valid structure be tempered
and given a cutting edge.

*

In contrast to the hard and impermeable shell of
the little ego (or the personality), the living skin of
the inner man that is comformable to Essential Being
and at the same time adapted to the world must become transparent, i.e., as well as being permeable to
Essential Being it must also be so to the repeated
deaths of the ego. This inner coating is the means by
which a man increases in strength and form and is
able to contain within himself the plentitude and unifying power that comes from Divine Being. It
should be remembered, however, that this "living
skin" remains alive only when, both in thought and
in action, man repeatedly risks the form that has
arisen during some precious moment of release from
the ego and inward union with the Ground. At such
times, the temptation to draw apart into some ideal
state of quiescence is very great. But by succumbing
to this temptation a man inevitably relapses into his
former condition.

It is impossible for Essential Being, thus enshrined
and, as it were, protected from the world, to give
light to or become creative in the world. Only when
the personality has become transparent is Essential
Being able to pursue its redemptive processes and
pierce with its rays the shell of the world-ego.
Therefore, he who has woken to Essential Being
fulfils his service to Divine Being by the way in

which he does "the one thing necessary": that is to say, by manifesting the Divine in the midst of the world in all his striving, all his creativity and all his love.

APPENDIX

Terminology

It should be understood for the purposes of this book that we distinguish between the conditioned world of manifestation, i.e., the reality of time and space, and that which is wholly unconditioned—the reality of *Divine Being*. This unconditioned reality is, in effect, the essence of all that exists. The manifestation of any living thing is inevitably the result of the integration of two poles, the one representing that which is conditioned by time and space, and the other Divine Being.

Essential being, again for the purposes of this book, is the term used in respect to the manner in which Divine Being is present in, say, a flower, an animal or a man. It is the individual form that Divine Being takes in any particular manifestation of life. In man, once he becomes aware of its presence, it is experienced as an inner image and at the same time as an inner path for him to follow. This essential being of man, representing as it does the presence of Divine Being—which continually strives to manifest itself in and through him—is at one and the same time partly hidden and partly revealed by what we call the *world-ego*. It is man's destiny so to transform himself that his essential being and, as a corollary, Divine

Being is able to manifest itself in its entirety in the world.

In so far as such transformation is achieved a man becomes a *Person,* which means to say a living form through which Divine Being may sound (*personare*).

Personality here indicates that aspect of man which enables him to measure up to the requirements of the world in a way that accords with the world's traditions and values.

The concept of the *Persona* can be referred back to C. G. Jung. It represents the façade which, to a greater or lesser degree, corresponds to that image of himself which a man fashions in order to represent the particular role he would like to play in the world. In his efforts toward this it invariably happens that he suppresses and thrusts into the unconscious everything that appears to be antagonistic to this image. Thus the *Shadow* is created.